Physics of the Atmosphere

Physics of the Atmosphere

Rodrigo Caballero

Department of Meteorology, Stockholm University, Sweden

IOP Publishing, Bristol, UK

ISBN 978-0-7503-1052-9 (ebook)
ISBN 978-0-7503-1053-6 (print)
ISBN 978-0-7503-1110-6 (mobi)

DOI 10.1088/978-0-7503-1052-9

Version: 20141101

IOP Expanding Physics
ISSN 2053-2563 (online)
ISSN 2054-7315 (print)

British Library Cataloguing-in-Publication Data: A catalogue record for this book is available from the British Library.

Published by IOP Publishing, wholly owned by The Institute of Physics, London

IOP Publishing, Temple Circus, Temple Way, Bristol, BS1 6HG, UK

US Office: IOP Publishing, Inc., 190 North Independence Mall West, Suite 601, Philadelphia, PA 19106, USA

Contents

Preface

This book derives from the lecture notes for a course in physical meteorology I taught at University College Dublin, Ireland, between the years 2005–2010. The course was the introductory step in a masters-level programme in meteorology, and was given in parallel with a course in dynamical meteorology. (The separation into 'physics' and 'dynamics' may sound strange to the uninitiated, but these are only somewhat unfortunate labels denoting what is in fact a natural split between the study of large-scale—$O(1000\,\mathrm{km})$—atmospheric motion and of the much smaller-scale processes such as radiation and cloud formation embedded within it.) The students coming into the programme varied greatly both in their background and aspirations: some were beginning PhD students while others aimed for a more vocational career in meteorology, and the only entrance requirement was familiarity with calculus and classical mechanics. The course therefore needed to be elementary and self-contained, but also rigorous and advanced enough to serve the research students while providing an introduction to the practical arcana of meteorology useful for the practitioners. In trying to achieve this balance I drew much inspiration from Bohren and Albrecht's excellent *Atmospheric Thermodynamics*, whose influence on chapters 2 and 3 will be obvious. The chapter on radiation owes much to G W Petty's *A First Course in Atmospheric Radiation* and also to R T Pierrehumbert's monumental *Principles of Planetary Climate*. The book has benefited from the comments of many colleagues and students over the years; I would particularly like to thank Isaac Held, Matt Huber, Adam Sobel and Johan Swanljung.

Rodrigo Caballero
rodrigo.caballero@misu.su.se
Stockholm, August 2014

Author biography

Rodrigo Caballero

Rodrigo Caballero is a Professor in the Department of Meteorology at Stockholm University. His research focuses on the role of the atmosphere within the climate system in past, present and future planetary climates.

IOP Publishing

Physics of the Atmosphere

Rodrigo Caballero

Chapter 1

Introduction

1.1 Composition of the atmosphere

Atmospheric constituents can be classed into two groups: those that are *well mixed* and those that are *variable*. Nitrogen and oxygen, which account for roughly 99% of the atmosphere (table 1.1), are examples of the former. These constituents have long residence times—they enter and leave the atmosphere much more slowly than the typical time it takes for turbulence to mix them up thoroughly throughout the atmosphere. As a result, their number fraction is essentially constant in space and time. Because of this, it is possible (from the physical and thermodynamic point of view) to treat the 'soup' of well-mixed gases as if it consisted of a single species with molecular weight equal to the constituents' *average* molecular weight. This fictitious gas is called *dry air*.

Table 1.1. Composition of the atmosphere. Number fractions (other than for water vapour) are specified with respect to dry air.

Constituent	Number fraction (%)
Nitrogen (N_2)	78.08
Oxygen (O_2)	20.95
Argon (Ar)	0.93
Carbon dioxide (CO_2)	0.038
Neon (Ne)	0.001818
Helium (He)	0.000524
Methane (CH_4)	0.0001745
Krypton (Kr)	0.000114
Hydrogen (H_2)	0.000055
Water vapour (H_2O)	0–5
Ozone (O_3)	0–0.00001

The most important variable constituent is water vapour. Water is the only atmospheric constituent that can change phase at the typical pressures and temperatures experienced in the Earth's atmosphere. It can condense to form clouds and precipitate out as rain, and it can evaporate from the surface and from cloud and rain droplets. These are fast processes, so the residence time of water vapour is brief. At any given instant, water vapour can account for anything between 5% of the atmosphere (near the surface in the tropics) and almost zero (in the stratosphere). To an excellent approximation, the atmosphere can be considered as a two-component gas, made up of variable proportions of dry air and water vapour.

1.2 Observed vertical structure

Vertical soundings of temperature, pressure and humidity are taken daily at a large number of meteorological stations spanning the globe. The data thus obtained are among the principal inputs to weather forecasting. There is a single sounding station in the Republic of Ireland, picturesquely sited on Valentia Island off the Kerry coast. Figure 1.1 shows a randomly selected sample sounding from Valentia. This serves to illustrate some key general features of the atmosphere's vertical structure.

- Pressure decreases smoothly with height. Surface pressure is about 1000 hPa.
- Temperature also decreases with height, though there is much more structure (more wiggles) in the profile. The rate of decrease or *lapse rate* is on average 6 °C km^{-1} (though in this particular sounding is closer to 5 °C km^{-1}). Above a certain height (about 15 km) the temperature *increases* with height (the lapse rate is negative). The cross-over point is known as the *tropopause*, separating the *troposphere* below from the *stratosphere* above. Surface temperature is about 15 °C or 288 K. A useful round-number value to keep in mind as a typical surface temperature is 300 K.
- Density decreases with height, mirroring the pressure. The surface value is roughly 1 kg m^{-3}.
- Humidity decreases sharply with height, dropping to near zero above 2–3 km.

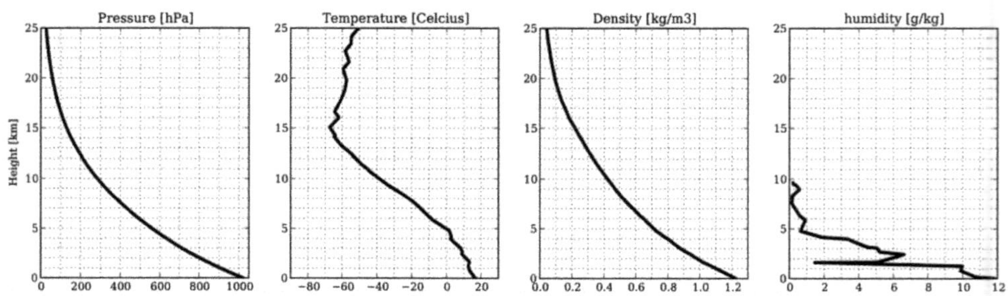

Figure 1.1. Sounding at Valentia station, south-west Ireland, at 00Z on 14 September 2005.

Chapter 2

Thermodynamics of dry air

2.1 Pressure, temperature and the ideal gas law

2.1.1 A simple derivation of the ideal gas law

The *pressure* exerted by a gas on a surface is due to the incessant pelting of the surface by the gas's constituent molecules. The more molecules there are near the surface, and the faster they are moving, the higher the pressure. Note that pressure is only due to the *local* properties of the gas and not to anything going on far away. Also, if the molecular motion has no preferred direction (i.e. if the gas as a whole is at rest), then pressure does not depend on the orientation of the surface. The number of molecules can be expressed as a *density* and, as we will see, the mean speed can be expressed as a *temperature*. Thus, pressure increases with density and temperature; the *ideal gas law* is a formal (i.e. mathematical) statement of this relationship.

To derive the gas law, we first need to quantify the force exerted by a molecule colliding with a surface. Consider a particle of mass m and velocity $\mathbf{v}$ colliding elastically with a wall over the course of the time interval t (figure 2.1). Using Newton's second law

$$\mathbf{F} = m\mathbf{a} = m\frac{d\mathbf{v}}{dt}, \tag{2.1}$$

we can write the cumulative force experienced by the particle over time t, a quantity known as the *impulse*, as

$$\mathbf{I} = \int_0^t \mathbf{F}dt = m(\mathbf{v}' - \mathbf{v}). \tag{2.2}$$

Assuming that the collision is elastic, the components of the impulse are

$$I_x = -2mv_x \tag{2.3}$$

$$I_y = 0. \tag{2.4}$$

doi:10.1088/978-0-7503-1052-9ch2

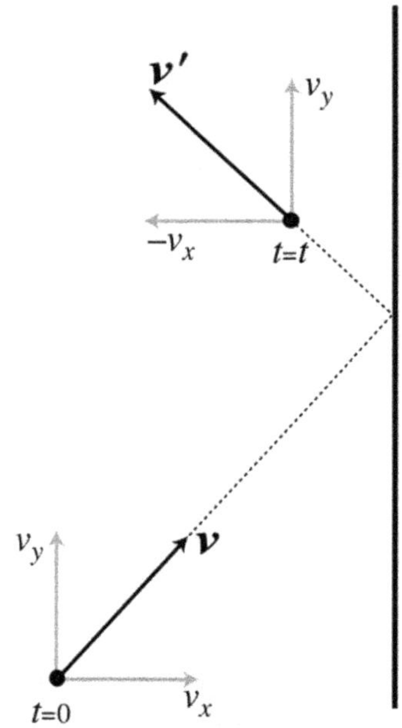

Figure 2.1. A particle bouncing elastically off a solid wall.

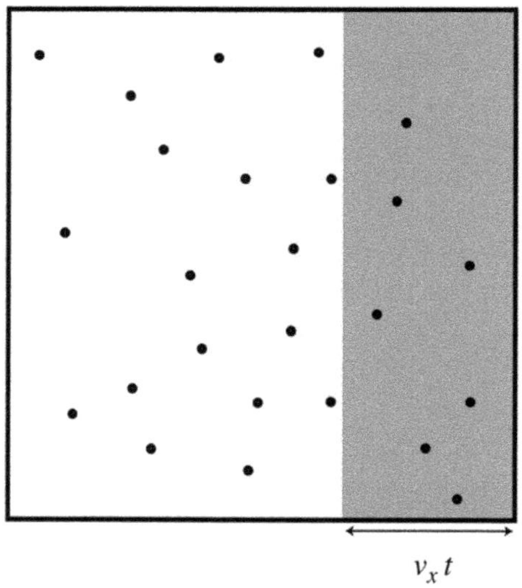

Figure 2.2. A box full of moving particles. All particles with x-velocity component v_x contained in the shaded volume will collide with the right-hand wall over the time period t.

This is the impulse on the particle; by Newton's third law, the impulse on the wall will be equal and opposite, giving $I_x = 2mv_x$.

Now consider a box of volume V containing N identical molecules (figure 2.2). We make four assumptions: (i) the particles have negligible volume; (ii) the particles do not attract each other (though they may collide); (iii) collisions with the wall are elastic; and (iv) the motion is *isotropic*, i.e. there are equal numbers of particles moving in each direction. The impulse on a wall of the box, orthogonal to the x-axis, due to particles moving with x-velocity v_x is

$$I_x(t, v_x) = 2mv_x \cdot Av_x t\frac{N}{V}P(v_x) = 2mnv_x^2 P(v_x)At \tag{2.5}$$

where A is the area of the wall, $n = N/V$ is the particle number density, and $P(v_x)$ is the probability that a given particle has x-velocity v_x. The formula above is derived by noting that all molecules having speed v_x and contained in a sub-volume of thickness $v_x t$ adjacent to the wall, will collide with the wall over the time period t.

To obtain the total impulse, we need to sum over the contributions due to all *positive* values of v_x (particles with negative v_x will obviously not collide with the wall):

$$\int_0^\infty I_x(t, v_x)\mathrm{d}v_x = mn\left\langle v_x^2\right\rangle At, \tag{2.6}$$

where we have defined

$$\left\langle v_x^2\right\rangle \equiv \int_{-\infty}^\infty v_x^2 P(v_x)\mathrm{d}v_x = 2\int_0^\infty v_x^2 P(v_x)\mathrm{d}v_x \tag{2.7}$$

and the second equality follows from the assumption of isotropy. Isotropy also implies that

$$\langle v^2\rangle = \left\langle v_x^2\right\rangle + \left\langle v_y^2\right\rangle + \left\langle v_z^2\right\rangle = 3\left\langle v_x^2\right\rangle \tag{2.8}$$

where for simplicity we use v^2 to denote the squared modulus $|\mathbf{v}|^2$. Finally, we define pressure as the impulse per unit time (i.e. force) per unit area:

$$p = \frac{\int_0^\infty I_x(t, v_x)\mathrm{d}v_x}{At} = \frac{1}{3}nm\langle v^2\rangle. \tag{2.9}$$

Thus, as anticipated above, pressure is proportional to density and to the mean squared velocity. To bring temperature into the picture, we simply define

$$kT \equiv \frac{1}{3}m\langle v^2\rangle, \tag{2.10}$$

where T is temperature (in kelvins) and $k = 1.38 \times 10^{-23}\,\mathrm{J\,K^{-1}}$ is *Boltzmann's constant*. Note that Boltzmann's constant is not a universal constant (like the speed of light), but is simply a unit conversion factor. *Temperature is just another name for the mean kinetic energy (KE) density of molecular motion.*

With this definition, we arrive at the *ideal gas law*:

$$p = nkT, \tag{2.11}$$

which can also be written in the more familiar form

$$p = \rho RT \tag{2.12a}$$

where $\rho = nm$ is the mass density and $R = k/m$ is the *gas constant*. Dry air has $m = 28.97$ atomic mass units (AMU)[1] and $R = 287\ \mathrm{J\,K^{-1}\,kg^{-1}}$.

Another way to write the ideal gas law is to multiply and divide (2.11) by Avogadro's number $N_A = 6.022 \times 10^{23}$ yielding

$$p = \mathfrak{n}R^*T \tag{2.12b}$$

where $\mathfrak{n} = n/N_A$ is the number of moles per unit volume and $R^* = kN_A = 8.31\ \mathrm{J\,K^{-1}\,mol^{-1}}$ is the *universal gas constant*.

2.1.2 Momentum flux and the pressure gradient force

The attentive reader will have noticed a considerable flaw in the previous section: in deriving the ideal gas law, pressure was defined only at a rigid wall. We would of course also like to define pressure in the interior of the gas, away from the wall. In fact, essentially all motion in the atmosphere arises through pressure *differences* from point to point, which accelerate the fluid in the direction of the smaller pressure (that is, down the pressure gradient). This is often conceptualized in terms of air parcels pressing their neighbours into motion. But the walls bounding an air parcel are purely imaginary, so when molecules from inside the parcel approach its edge they do not bounce back but pass through unimpeded. So how exactly does an air parcel exert pressure on its neighbours?

Part of the answer is that the molecules entering an air parcel from outside bring into the parcel a net amount of momentum directed normal to the parcel's wall and into the parcel (which we can define to be the positive direction). Within a short distance, these foreign particles collide with the parcel's particles and effectively cede their positive momentum to the parcel. This is only half the story, however; the other half is that particles exiting the parcel take with them a similar amount of momentum but directed *out* of the parcel—that is, they remove negative momentum, which is the same as adding positive momentum.

To formalize these ideas, we need to define a flux of momentum. Given any physically measurable quantity (such as mass, particle number or momentum), its *flux* is the amount of that quantity flowing through a unit area per unit of time. For instance, consider a pipe of cross-sectional area A containing an ideal gas with an isotropic velocity distribution (figure 2.3). Following the same reasoning used to obtain (2.5) and (2.6), we can write the flux of particles across an imaginary cross-sectional surface (indicated by S in the figure) as

$$\frac{1}{At}\int_0^\infty \left[v_x tAnP(v_x) - v_x tAnP(-v_x)\right]\mathrm{d}v_x = 0, \tag{2.13}$$

where the first term on the left-hand side counts particles moving from left to right while the second term counts those moving from right to left. Because of isotropy, $P(v_x) = P(-v_x)$ and so the result is 0. There is no net transfer of particles across the imaginary surface: though its constituent molecules are whizzing around tirelessly, the gas as a whole is at rest (there is no wind in this wind tunnel).

[1] The atomic mass unit, 1.661×10^{-27} kg, is defined as 1/12 the mass of a ^{12}C atom.

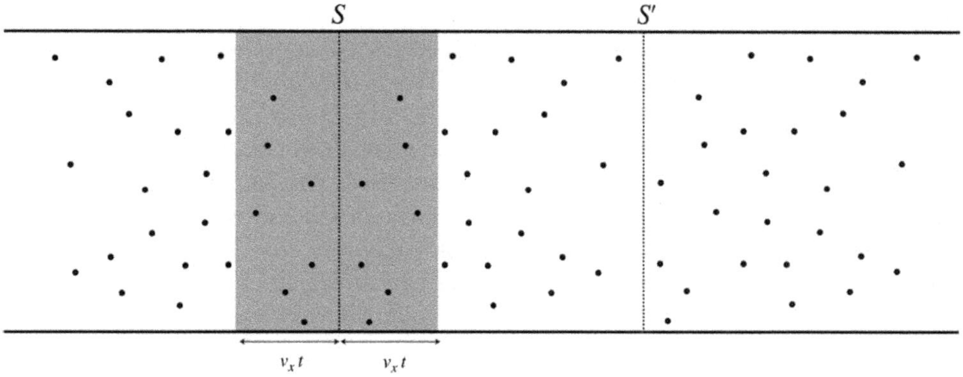

Figure 2.3. A pipe full of moving particles. All particles with x-velocity component $\pm v_x$ contained in the shaded volume will pass through imaginary surface S in time t.

Now consider the flux of x-momentum across the same imaginary surface:

$$\frac{1}{At}\int_0^{\infty}\left[mv_x^2 tAnP(v_x) + mv_x^2 tAnP(-v_x)\right]\mathrm{d}v_x = mn\left\langle v_x^2\right\rangle = p. \qquad (2.14)$$

Though there is no net flux of particles, there is a positive flux of momentum, which is exactly equal to the pressure that would be exerted on a solid wall. *Pressure is a flux of momentum—more specifically, it is the flux of the momentum component normal to a given surface.*

Consider what happens when pressure is not uniform in the pipe. Let's say the molecules are moving a bit more slowly at position S' in the pipe, so that there the pressure p' is a bit less than at S. The molecular motion is still isotropic, so there is still no overall flow in the pipe. But now, more momentum is entering through S than is leaving through S'. The net momentum of the gas contained in the segment SS' is increasing: if it was zero initially, it will grow and become non-zero. In other words, the gas will accelerate and begin to flow. The flow occurs *down* the pressure gradient, i.e. from high pressure to low.

2.1.3 The Maxwell–Boltzmann distribution

In the derivation of the ideal gas law, the probability distribution $P(v_x)$ played a key role. In particular, we defined the temperature as a quantity proportional to the mean squared speed of the molecules. But just what is this distribution? Can it change over time?

The possibility of time variability causes a bit of a problem. Consider, for instance, a gas in an isolated box. Imagine you initialize all the molecular velocities to point in the same direction, like a beam. There is a well-defined velocity distribution, and so we can (at least in principle) define a temperature. However, molecular motion is a chaotic process. Any tiny initial anomalies in the molecules' direction will be amplified as molecules collide with walls and with each other. This will remove molecules from the original beam and create a set of 'diffuse' molecules moving in random directions. The probability distribution will thus change; as a result, the temperature and pressure will also, in general, change over time. But this contradicts our usual experience (or assumption) that a gas left alone in an isolated box will maintain constant temperature and pressure. This is the definition of *thermodynamic equilibrium* and classical thermodynamics deals only with equilibrium systems.

So can we conceive of a velocity distribution which does not change in time, which would characterize a gas in equilibrium? The answer is yes. The key characteristic of such an *invariant* velocity distribution is that collisions will redirect as many molecules *away* from a given velocity as they do *into* that same velocity. This condition can be expressed mathematically as follows. Consider two molecules, moving with velocities $\mathbf{v}_1$ and $\mathbf{v}_2$ respectively, which after colliding take on velocities $\mathbf{v}'_1$ and $\mathbf{v}'_2$. Then a necessary and sufficient condition for the velocity distribution f to be time-invariant is

$$f(\mathbf{v}_1)f(\mathbf{v}_2) = f(\mathbf{v}'_1)f(\mathbf{v}'_2) \tag{2.15}$$

for all possible sets of initial and final velocities. Note that there is no need for the collisions themselves to be isotropic—i.e. some collisions may be more frequent than others.

Given the condition above, we can derive the full functional form of the invariant distribution. Taking the logarithm of equation (2.15),

$$\ln f(\mathbf{v}_1) + \ln f(\mathbf{v}_2) = \ln f(\mathbf{v}'_1) + \ln f(\mathbf{v}'_2), \tag{2.16}$$

which implies that $\ln f(\mathbf{v})$ must be a *constant of motion*, i.e. a quantity that is conserved at all times during the motion. A simple point particle has four constants of motion: the KE $mv^2/2$ and the three components of the linear momentum, mv_x, mv_y, mv_z. If the mass is constant and equal for all particles, then the relevant constants of motion are v^2, v_x, v_y and v_z. Equation (2.16) implies that $\ln f(\mathbf{v})$ must be a linear combination of these constants of motion, which can be written in the most general form as

$$\ln f(\mathbf{v}) = -B|\mathbf{v} - \mathbf{v}_0|^2 + \ln A, \tag{2.17}$$

implying

$$f(\mathbf{v}) = A\, \mathrm{e}^{-B|\mathbf{v}-\mathbf{v}_0|^2} \tag{2.18}$$

where A, C and the three components of $\mathbf{v}_0$ are five constants. To determine these constants, we first note that the mean velocity

$$\langle \mathbf{v} \rangle = A\int \mathbf{v}\mathrm{e}^{-B|\mathbf{v}-\mathbf{v}_0|^2}\mathrm{d}v_x\mathrm{d}v_y\mathrm{d}v_z = A\int (\mathbf{v} - \mathbf{v}_0)\mathrm{e}^{-Bv^2}\mathrm{d}v_x\mathrm{d}v_y\mathrm{d}v_z = \mathbf{v}_0, \tag{2.19}$$

so if we assume that the gas has no mean motion, then $\mathbf{v}_0 = 0$. We can then use spherical coordinates to write

$$f(v)\mathrm{d}v_x\mathrm{d}v_y\mathrm{d}v_z = A\mathrm{e}^{-Bv^2}\, v^2 \cos\varphi\ \mathrm{d}\varphi\mathrm{d}\theta\mathrm{d}v \tag{2.20}$$

and integrating over φ and θ we find that the probability density function for the speed v is

$$f(v) = 4\pi A\, v^2\, \mathrm{e}^{-Bv^2}. \tag{2.21}$$

The constants A and B are determined by the normalization condition

$$\int_0^\infty f(v)\mathrm{d}v = 1 \tag{2.22}$$

and the definition of temperature

$$kT = \frac{1}{3}m\langle v^2\rangle = \frac{1}{3}m\int_0^\infty v^2 f(v)\mathrm{d}v. \tag{2.23}$$

Carrying out the integrations and solving for A and B yields the *Maxwell–Boltzmann distribution*

$$f(v) = 4\pi\left(\frac{m}{2\pi kT}\right)^{3/2} v^2\, \mathrm{e}^{-\frac{mv^2}{2kT}}. \tag{2.24}$$

This is the probability distribution function for the molecular speeds of a gas in equilibrium, in terms of the single macroscopically measurable quantity T. Note that the distribution is isotropic, i.e. does not depend on the direction but only on the speed of the molecules. It can be shown that a gas initialized with any velocity distribution will always evolve over time to assume just such a time-invariant isotropic distribution, a process referred to as *relaxation to thermodynamic equilibrium*.

2.2 Hydrostatic balance

2.2.1 The momentum budget of a stationary layer of air subject to gravity

Now let's look at the particles in the box again, but this time we'll add the effect of gravity. Consider a thin horizontal layer of thickness Δz, as shown in figure 2.4. If we assume that the gas as a whole is at rest (or is moving with constant velocity), then the layer cannot be gaining or losing momentum. Recalling from the previous section that pressure is simply the flux of momentum, we can write the vertical momentum budget for the layer as

$$p(z)A - p(z+\Delta z)A - A\Delta z nmg = 0 \tag{2.25}$$

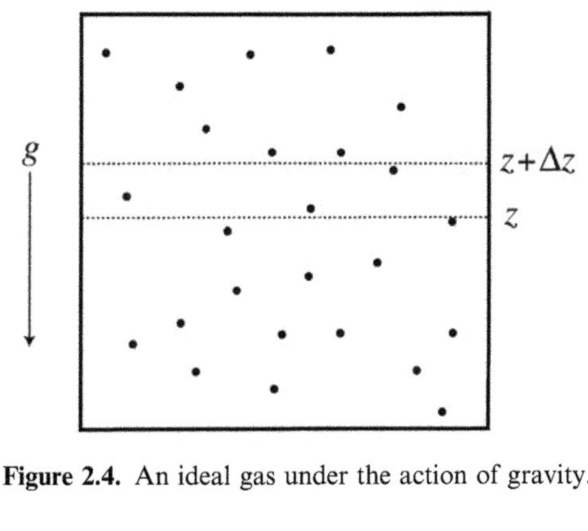

Figure 2.4. An ideal gas under the action of gravity.

which can be rearranged to

$$\frac{p(z + \Delta z) - p(z)}{\Delta z} = -nmg \tag{2.26}$$

and in the limit $\Delta z \to 0$:

$$\frac{\mathrm{d}p}{\mathrm{d}z} = -\rho g, \tag{2.27}$$

which is the *equation of hydrostatic balance*. Remember the key assumption: the gas is not accelerating in the vertical (though it may be moving and accelerating in the horizontal).

2.2.2 The hydrostatic approximation

The ideal gas law (2.12) is a single relationship between the three thermodynamic variables p, ρ and T, and puts limited constraints on atmospheric structure: there are infinitely many combinations of T and ρ that give the same p. Hydrostatic balance gives an extra relationship among the variables and thus helps narrow the range of possibilities—in a hydrostatically balanced atmosphere we need only specify the temperature distribution, say, and the other two fields are automatically determined. Thinking about an atmosphere in hydrostatic balance is therefore much easier than thinking about one that is not.

The real atmosphere is of course in perpetual, chaotic motion, and this includes time-varying vertical motion, which implies vertical acceleration. The vertical momentum equation then takes the form

$$\rho\frac{\mathrm{d}^2 z}{\mathrm{d}t^2} = -\frac{\mathrm{d}p}{\mathrm{d}z} - \rho g, \tag{2.28}$$

so that strictly speaking the atmosphere is never in precise hydrostatic balance. But infinite precision is for fanatics; the more productive approach is to look for simplifications which are true at a level of precision *good enough for the problem at hand.* In our case, the relevant question is whether the atmosphere is close enough to hydrostatic balance for the purposes for which we need it in this book.

Consider an atmosphere with three-dimensional, time-varying pressure and mass distributions $p(x, y, z, t)$ and $\rho(x, y, z, t)$. Now let's take the horizontal and time average of the mass distribution and call it $\rho_0(z)$, and define a pressure distribution $p_0(z)$ which exactly balances the mean mass distribution:

$$\frac{\mathrm{d}p_0}{\mathrm{d}z} = -\rho_0 g. \tag{2.29}$$

We can then write

$$p(x, y, z, t) = p_0(z) + p'(x, y, z, t), \quad \rho(x, y, z, t) = \rho_0(z) + \rho'(x, y, z, t), \tag{2.30}$$

where p' is simply what's left over when subtracting p_0 from p (and analogously for ρ'). When performing this kind of partitioning, p_0 and ρ_0 are usually called the *basic state* and p', ρ' the *perturbation*.

In this book, the typical situation in which we invoke the hydrostatic approximation is when looking at the behaviour of a small 'test parcel' of air displaced vertically within a larger-scale environment of known properties: we would like to relate the thermodynamic state of the parcel to that of the surrounding air, and it helps if the surroundings can be assumed to be in hydrostatic balance. In this case, all we require is that

$$p' \ll p_0 \quad \text{and} \quad \rho' \ll \rho_0 \tag{2.31}$$

for the surroundings; i.e. we want the perturbation field to be negligibly small so that we can focus on the basic state (which is hydrostatic by definition). Under what conditions is (2.31) true?

Substituting (2.30) in (2.28) we can write with no approximation

$$\left(\rho_0 + \rho'\right)\frac{\mathrm{d}^2 z}{\mathrm{d}t^2} = -\frac{\mathrm{d}p'}{\mathrm{d}z} - \rho' g. \tag{2.32}$$

In order for this equation to be true, at least two of the terms must be of the same order of magnitude at all times (the remaining term can be of the same magnitude or smaller, but not larger). Suppose in particular that the left-hand side is mostly balanced by the second term on the right-hand side, so that

$$\frac{\left|\mathrm{d}^2 z/\mathrm{d}t^2\right|}{g} \sim \frac{\rho'}{\rho_0 + \rho'} \tag{2.33}$$

where the symbol ~ means 'has the same order of magnitude as'. It is then clear that (2.31) will be satisfied if

$$\frac{\left|\mathrm{d}^2 z/\mathrm{d}t^2\right|}{g} \ll 1, \tag{2.34}$$

that is if vertical accelerations are much smaller than g.

Is (2.34) typically true in the atmosphere? To answer this question we need to look at the mechanisms which generate vertical acceleration. A common one is *horizontal convergence*: when the horizontal wind field is not uniform, air tends to bunch up in some areas (see figure 2.5). Since air behaves essentially as an incompressible fluid in this context, air must accelerate vertically in these convergence zones. To estimate

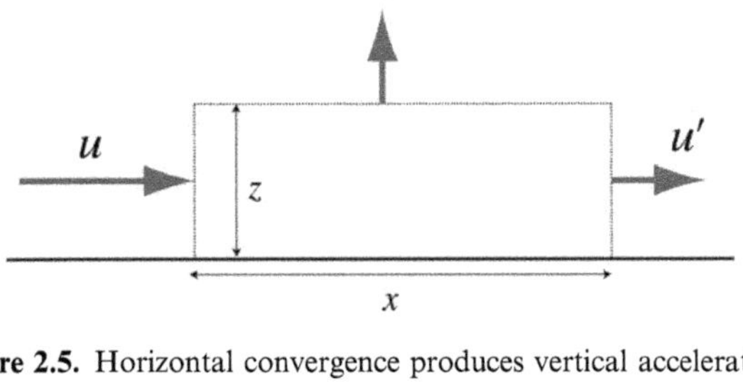

Figure 2.5. Horizontal convergence produces vertical acceleration.

the size of the acceleration, note that the volume of the box sketched in figure 2.5 is xz and must stay constant in time because of incompressibility. Thus

$$\frac{\mathrm{d}}{\mathrm{d}t}(xz) = 0 = x\frac{\mathrm{d}z}{\mathrm{d}t} + z\frac{\mathrm{d}x}{\mathrm{d}t} = xw - z\Delta u, \tag{2.35}$$

where $\Delta u = u - u'$, and so

$$w = \frac{z}{x}\Delta u. \tag{2.36}$$

The vertical acceleration is then

$$\frac{\mathrm{d}^2 z}{\mathrm{d}t^2} = \frac{w}{x}\Delta u + \frac{z}{x^2}\Delta u^2 = 2\frac{z}{x^2}\Delta u^2 \tag{2.37}$$

where we have taken Δu to be constant.

For an order-of-magnitude estimate, we take $z \sim H$, $x \sim L$ and $\Delta u \sim U$ so that

$$\frac{\left|\mathrm{d}^2 z/\mathrm{d}t^2\right|}{g} \sim \left(\frac{H}{L}\right)^2 \left(\frac{U}{\sqrt{gH}}\right)^2 \tag{2.38}$$

H/L is known as the *aspect ratio* and achieves its greatest values in regions such as *surface fronts* where $H \sim L \sim 1$ km. Taking $U \sim 10\,\mathrm{m\,s^{-1}}$ then makes $|\mathrm{d}^2 z/\mathrm{d}t^2|/g \sim 0.01$, so the hydrostatic approximation is valid. For larger-scale motion, L ranges from 10 to 10^4 km but H is at most 10 km, so the hydrostatic approximation becomes even better.

Note that (2.33) implies that the pressure and mass perturbations are *not* in hydrostatic balance. As we argued above, this is not of concern for the questions addressed in this book. It would become a problem, however, if this book dealt with atmospheric dynamics—specifically, with horizontal motion. The prime driver of the horizontal wind is the horizontal pressure gradient, which is associated entirely with p' since p_0 is horizontally uniform. It greatly simplifies the study of dynamics if p' can be computed from ρ' using the hydrostatic relation. In order for the perturbation to be hydrostatically balanced we require the stronger condition

$$\frac{\left|\mathrm{d}^2 z/\mathrm{d}t^2\right|}{g} \ll \frac{\rho'}{\rho_0}. \tag{2.39}$$

To make further progress in evaluating this condition, we need to introduce a number of new concepts. We will take up this question again in section 2.5.3.

2.2.3 Surface pressure and the mass of the atmosphere

If we integrate (2.27) vertically from the surface ($z = 0$) up to a very large height, where $p = 0$, we find

$$p_\mathrm{s} = \int_0^\infty g\rho\,\mathrm{d}z = Mg, \tag{2.40}$$

where p_s is surface pressure and M is the total mass of atmosphere contained in a column of unit cross section. Thus, *for a fluid in hydrostatic equilibrium*, pressure at a certain point is proportional to the mass of fluid above that point.

2.2.4 Scale height

Since g and ρ are always positive, hydrostatic equilibrium (2.27) implies that *pressure always decreases with height*. To compute the vertical structure of pressure, we need to know the behaviour of ρ with height, which depends on temperature: using the ideal gas law, we have

$$\frac{\mathrm{d}p}{\mathrm{d}z} = -\frac{g}{RT}p. \tag{2.41}$$

This can be integrated to give

$$p(z) = p_0 \mathrm{e}^{-z/H(z)} \tag{2.42}$$

where p_0 is surface pressure and we have introduced the *scale height*

$$H(z) = \frac{R\langle T\rangle}{g}, \tag{2.43}$$

where

$$\langle T\rangle = \left(\frac{1}{z}\int_0^z \frac{\mathrm{d}z}{T}\right)^{-1} \tag{2.44}$$

is the so-called *harmonic mean* of temperature between the ground and z.

Note that the scale height thus defined depends on z, since T generally changes with height. In our atmosphere, however, its variation is actually not so great (see figure 2.6), so it's not too bad an approximation to take H as a constant. In this case, p decreases exponentially with a decay rate H. As shown in figure 2.6, the fit is quite good.

2.2.5 Layer thickness and the hypsometric equation

Equation (2.42) implies that as temperature increases, pressure decreases more slowly with height, i.e. the atmosphere becomes deeper. To show this explicitly, we can invert (2.42) to obtain the *hypsometric equation* (from the Greek *hypsos* for height):

$$z = \ln\left(\frac{p_0}{p}\right) H(p) \tag{2.45}$$

where $H(p)$ is the scale height for the layer between the ground and pressure-level p.

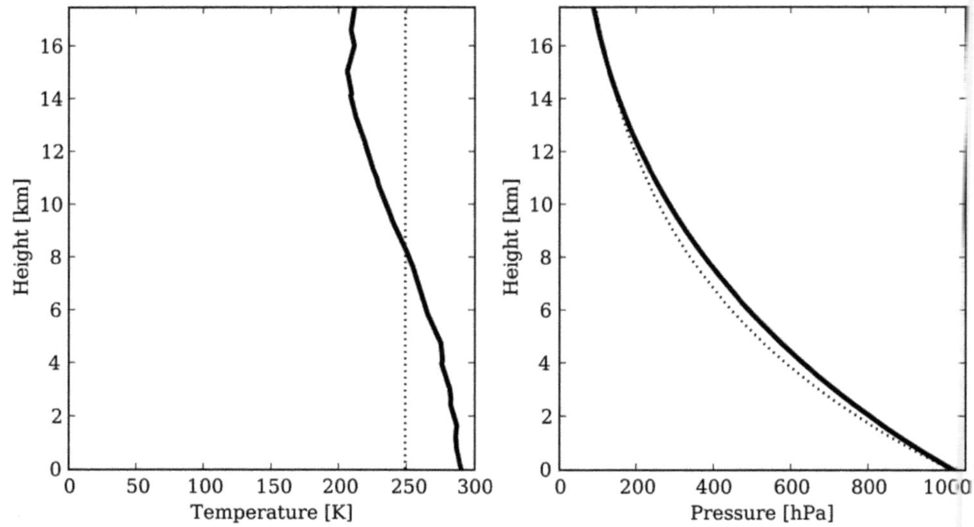

Figure 2.6. (Left panel) Temperature profile at Valentia. The profile does not depart from the vertical average (dotted line) by more than about 20%. (Right panel) Observed pressure profile (solid) and approximation computed using constant scale height (dotted).

Exercise 2.2.5: Show that the harmonic mean (2.44) can also be written as

$$\langle T \rangle = \frac{\int_p^{p_0} T \, \mathrm{d} \ln p}{\int_p^{p_0} \mathrm{d} \ln p}. \tag{2.45}$$

The hypsometric equation says that the thickness of an atmospheric layer bounded by any two pressures is proportional to the (harmonic) mean temperature within the layer. This implies that a horizontal temperature gradient will create a horizontal pressure gradient, generally leading to motion.

2.2.6 Surface pressure and sea level pressure

Atmospheric motions cause changes in pressure at the surface—this is why a barometer is useful in predicting weather. However, as figure 2.7 makes clear, these changes are tiny (~10 mb) compared to changes in surface pressure due to orography—mountain ranges on Earth have heights comparable to the scale height. To plot and track the meteorologically significant pressure perturbations, we need to subtract the orographic effect. This is done by projecting surface pressure over mountain ranges down to sea level using the hypsometric equation. In other words, we plot the sea-level pressure that would be obtained if the mountain were replaced by air. This requires an arbitrary assumption about the temperature of the air that replaces the mountain. The simplest assumption is that this temperature is the same as the air temperature just above the mountain. This is good for low mountains, but

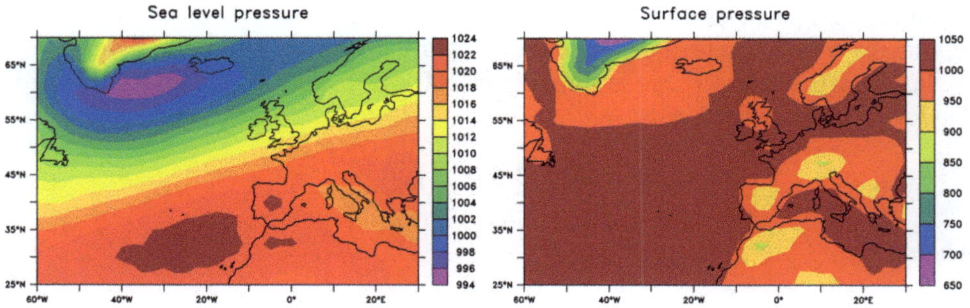

Figure 2.7. January climatology of sea level pressure (left) and surface pressure (right) as represented in the National Center for Environmental Prediction (NCEP) reanalysis product (Kalnay *et al* 1996).

for tall mountains it can lead to substantial biases. It is a particular problem over Greenland, which is tall (~3 km) and very cold.

2.3 Atmospheric energy and the first law of thermodynamics

2.3.1 Energy of a point mass in Earth's gravitational field

For a point of mass m interacting gravitationally with Earth (assumed to be spherical), Newton's laws give

$$-\frac{GM_{\mathrm{E}}m}{\left(R_{\mathrm{E}}+z\right)^2}=m\frac{\mathrm{d}v_z}{\mathrm{d}t}, \tag{2.47}$$

where G is the universal gravitational constant, M_{E} and R_{E} are Earth's mass and radius, and z is height above the surface. Multiplying by v_z,

$$-\frac{GM_{\mathrm{E}}m}{\left(R_{\mathrm{E}}+z\right)^2}\frac{\mathrm{d}z}{\mathrm{d}t}=mv_z\frac{\mathrm{d}v_z}{\mathrm{d}t}, \tag{2.48}$$

which can be re-written

$$\frac{\mathrm{d}}{\mathrm{d}t}\left(\frac{1}{2}mv_z^2-\frac{GM_{\mathrm{E}}m}{R_{\mathrm{E}}+z}\right)=0. \tag{2.49}$$

The quantity in parentheses is an invariant of the motion called *energy*. Note that the expression 'conservation of energy' is something of a tautology, since energy is conserved by definition. Because of the derivative on the left-hand side, we are free to add a constant to the definition of energy. We choose the constant $GM_{\mathrm{E}}m/R_{\mathrm{E}}$, which leads to energy defined as

$$E=\frac{1}{2}mv_z^2+\frac{GM_{\mathrm{E}}m}{R_{\mathrm{E}}}\left(1-\frac{1}{1+z/R_{\mathrm{E}}}\right). \tag{2.50}$$

The first term on the right-hand side is the *kinetic energy* (KE) and is always positive. The second term is the *gravitational potential energy* (PE) and is also positive assuming that particles do not go below Earth's surface. Because energy is constant, any increase in KE must be compensated by a decrease in PE and vice versa.

Since Earth's atmosphere is only about 100 km thick, $z \ll R_E$, and a first-order Taylor expansion gives

$$E = \frac{1}{2}mv_z^2 + mgz \tag{2.51}$$

where

$$g = \frac{GM_E}{R_E^2} \tag{2.52}$$

is the acceleration due to gravity at the Earth's surface.

2.3.2 Molecular interpretation of the scale height

Recalling that $R = k/m$, the scale height (2.43) can be written

$$H = \frac{k\langle T\rangle}{mg} = \frac{m\langle v^2\rangle/3}{mg} \tag{2.53}$$

which allows us to interpret H as roughly the distance a molecule must rise for its PE to match its typical KE. By conservation of energy, particles will be able to rise higher the greater the temperature or the smaller the gravity. Thus, an atmosphere in hydrostatic equilibrium gets thicker as it gets warmer, and for a given temperature, a smaller planet will have a thicker atmosphere.

2.3.3 Escape velocity and why we don't lose our atmosphere

Imagine a molecule near the top of the atmosphere which is moving directly upward. Since it is near the top of the atmosphere, it is unlikely to collide with other molecules. If its initial velocity is low, it will gradually convert all its KE into PE, come to a halt, and fall back down to Earth. However, (2.50) shows that it takes a *finite* amount of energy, $GM_E m/R_E$, to lift a mass from the ground to infinity. If the molecule's initial KE is greater than this value, it will never come to a halt and will escape the Earth for good. The threshold velocity is known as the *escape velocity*

$$v_e = \left(\frac{2GM_E}{R_E}\right)^{1/2} \tag{2.54}$$

and is about $11.2\ \mathrm{km\,s^{-1}}$.

If a substantial number of molecules achieve escape velocity, then the atmosphere will rapidly be lost to space. The probability that a given molecule's speed will exceed v_e is given by

$$f_e = \int_{v_e}^{\infty} f(v)\mathrm{d}v \tag{2.55}$$

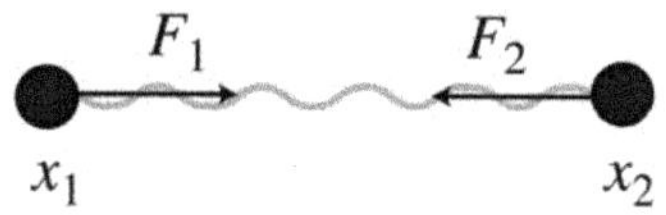

Figure 2.8. Two point masses joined by a spring.

where $f(v)$ is the Maxwell–Boltzmann distribution (2.24). There will always be *some* chance of finding a molecule moving faster than v_e. However, if m is large or T is small, the probability will be so small that it would take billions of years to lose the atmosphere. Given typical Earth temperatures, all atmospheric constituents find themselves in this situation.

2.3.4 The energy of two point masses joined by a spring

An *ideal spring* is one which exerts a force proportional to its extension. Consider two identical point masses joined by such a spring (figure 2.8). The equation of motion for mass 1 is

$$F_1 = K\Delta x = m\dot{v}_1 \tag{2.56}$$

where K is the spring constant, $\Delta x = x_2 - x_1$ and we are using the notation $\dot{x} = \mathrm{d}x/\mathrm{d}t$. For mass 2 we have

$$F_2 = -K\Delta x = m\dot{v}_2 \tag{2.57}$$

implying $\dot{v}_1 = -\dot{v}_2$. Subtracting one equation from the other and multiplying through by $v_2 - v_1$:

$$-2K\Delta\dot{x}\Delta x = m(v_2 - v_1)(\dot{v}_2 - \dot{v}_1) = 2m(v_1\dot{v}_1 + v_2\dot{v}_2) \tag{2.58}$$

which can be rewritten

$$\frac{\mathrm{d}}{\mathrm{d}t}\left(\frac{1}{2}mv_1^2 + \frac{1}{2}mv_2^2 + \frac{1}{2}K\Delta x^2\right) = 0. \tag{2.59}$$

Thus, this system also has an energy, which can be thought of as the sum of the KE of the two masses plus an elastic PE. This is a fairly good model for a diatomic molecule, except for one important detail: at the molecular level, quantum mechanics applies, and the energy of the molecule can only take on a discrete set of values. This has important macroscopic consequences, as we will see shortly.

2.3.5 External and internal energy

Let's consider now the energy of a large collection of particles—that is, let us look at energy from the macroscopic perspective. For definiteness, consider N diatomic molecules in a rigid, isolated box. 'Rigid' means that the molecules conserve their KE when colliding with the walls, and that the box does not change shape or size; 'isolated' means that no forces (other than gravity) act on the molecules. Under these circumstances, the total energy of the molecules will remain constant. The total energy of the molecules is simply the sum of their KE, gravitational PE and elastic

PE (we neglect electronic and other forms of energy since they are largely irrelevant for atmospheric processes):

$$E = \sum_i \left(\frac{1}{2} m v_i^2 + m g z_i + R_i + V_i \right) \tag{2.60}$$

where the sum extends over all molecules in the box, z_i and v_i are the height and speed of (the centre of mass of) each molecule, R_i represents rotational KE and V_i vibrational PE (the elastic PE due to stretching of the molecular bond).

Now let's write

$$v_i = \langle v \rangle + \tilde{v}_i, \tag{2.61}$$

that is as the sum of the mean velocity and a zero-sum deviation from the mean. Then

$$\sum_i v_i^2 = N\langle v^2 \rangle = N\langle v \rangle^2 + N\langle \tilde{v}^2 \rangle, \tag{2.62}$$

and the total energy can be written

$$E = \frac{N}{2} m\langle v \rangle^2 + Nmg\langle z \rangle + \frac{N}{2} m\langle \tilde{v}^2 \rangle + N\langle R \rangle + N\langle V \rangle. \tag{2.63}$$

The first two terms on the right-hand side are the KE and gravitational PE associated with the centre of mass of the N molecules: they can be thought of as the macroscopic (or *external*) KE and PE. The rest of the right-hand side is the energy associated with microscopic molecular motions around the centre of mass, and is referred to as the *internal* or *thermal* energy of the system, and given the symbol U.

The distinction between internal and external energy is central to thermodynamics. External energy is energy organized at the macroscopic level, which can be harnessed to do useful work in a machine. Internal energy is disorganized random motion which sums to zero. If we were to apply the categories of the Protestant work ethic to senseless molecules (an example of anthropomorphism, which is very hard to resist), we would call external energy 'good' and internal energy 'bad'. The science of thermodynamics arose in the 19th century when engineers were faced with the problem of converting the 'bad' energy released by burning coal to the 'good' energy of moving pistons.

2.3.6 Heat capacity

The name 'thermal energy' suggests a close association between this type of energy and temperature. For a monatomic gas in a rigid, isolated box, the internal energy has no rotational or vibrational components, and temperature *is* the internal energy:

$$U = \frac{N}{2} m\langle \tilde{v}_i^2 \rangle = \frac{3}{2} NkT. \tag{2.64}$$

Heat capacity measures the change in internal energy for a given temperature change. The heat capacity *at constant volume* is defined by the change in U when the volume of the system does not change:

$$C_V = \left.\frac{\partial U}{\partial T}\right|_V \tag{2.65}$$

and we can immediately see from (2.64) that for a monatomic gas,

$$C_V = \frac{3}{2}Nk. \tag{2.66}$$

When we consider polyatomic molecules, the relation between internal energy and temperature becomes more problematic, since we don't know how to express rotational and vibrational energy in terms of temperature. This problem is solved by the *principle of equipartition*, which states that *in equilibrium, each independent degree of freedom of the gas has an average energy of $kT/2$*. A 'degree of freedom' is a unique way for the system to move: for instance, motion of a point mass in two dimensions has two degrees of freedom, since any direction in the plane can be expressed as a sum of two vectors. The physical content of the principle is that, through molecular collisions, energy is quickly and evenly redistributed throughout the gas, so that each degree of freedom of each molecule has the same average energy as all the others.

A diatomic molecule moving in three dimensions has six degrees of freedom: three for translation of the centre of mass, two for rotation around the centre of mass and one for vibration (stretching along the molecular axis). We thus expect its heat capacity to be $6Nk/2$. Is this actually true? In fact, measurements of the heat capacity at constant volume of diatomic gases at typical atmospheric temperatures show $C_V \simeq 5Nk/2$, as if we had one less degree of freedom. The reason for this is quantum-mechanical. Molecules are not free to have any value of energy they want; they can only occupy discrete energy levels, with intermediate levels forbidden. For translational and rotational energy, the separation between energy levels is small compared with typical atmospheric values of $kT/2$, and so these degrees of freedom actively participate in the motion. Vibrational energy, on the other hand, has very widely separated energy levels: with the typical thermal energy available to atmospheric molecules, it is very rarely possible to cause vibrational transitions. As a result, vibrational degrees of freedom are effectively 'frozen out' and do not contribute to the heat capacity.

In general, a gas composed of molecules having s effective degrees of freedom has a heat capacity

$$C_V = \frac{s}{2}Nk. \tag{2.67}$$

Dividing through by the total mass of gas, Nm, we obtain the *specific heat capacity at constant volume*,

$$c_V = \frac{s}{2}R. \tag{2.68}$$

2.3.7 Heating, working and the first law of thermodynamics

Consider once more a set of molecules within a box. If we heat the gas by some means (radiative or conductive) at a rate Q, and the box is rigid, then the gas's temperature will increase at a rate set by $C_{\rm v}$:

$$\frac{\mathrm{d}U}{\mathrm{d}t} = \left.\frac{\partial U}{\partial T}\right|_{\rm v} \frac{\mathrm{d}T}{\mathrm{d}t} = C_{\rm v}\frac{\mathrm{d}T}{\mathrm{d}t} = Q. \qquad (2.69)$$

Since $p = nkT$ and the number density n is not changing, p will also increase.

But what if the volume of the box can change? Consider the set-up shown in figure 2.9. Initially, the pressure of the gas matches the weight of the piston, so that the system is in equilibrium. If we heat the gas without changing the volume, both temperature and pressure will increase as above. The piston will now be out of equilibrium and will move upwards. As it does so, the gas expands and the pressure decreases until equilibrium is re-established. Raising the piston increases its gravitational PE at the expense of the gas's internal energy. Following (2.69), this can be formalized as

$$C_{\rm v}\frac{\mathrm{d}T}{\mathrm{d}t} + W = Q. \qquad (2.70)$$

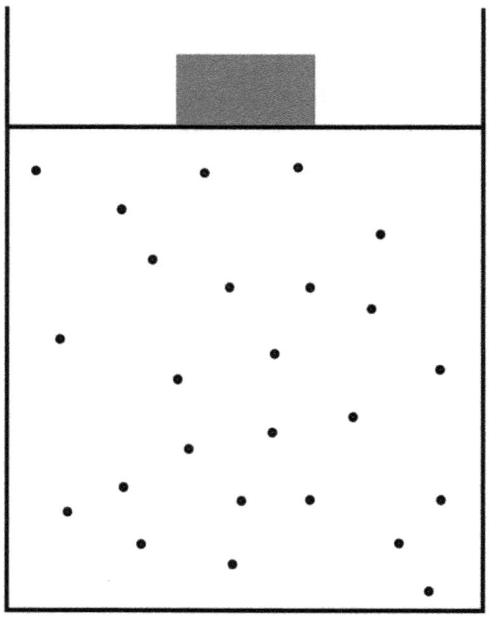

Figure 2.9. A cylinder fitted with a sliding piston. The pressure in the enclosure is kept constant at the value set by the weight of the piston.

where W is the rate at which work is done to lift the piston. If the piston rises a distance Δz in time Δt,

$$W = \frac{pA\Delta z}{\Delta t} \to p\frac{\mathrm{d}V}{\mathrm{d}t}. \tag{2.71}$$

Dividing through by the total mass of the gas, Nm, we have

$$c_\mathrm{v}\frac{\mathrm{d}T}{\mathrm{d}t} + p\frac{\mathrm{d}\alpha}{\mathrm{d}t} = J \tag{2.72}$$

$\alpha = 1/\rho$ is the specific volume and $J = Q/Nm$ is the heating rate per unit mass. The physical content of (2.72) is quite simple: if a gas is heated, the energy added will be used both to increase the temperature (i.e. increase the mean speed of molecular motion) and to expand the gas doing work on the environment. This equation is usually referred to as the 'law of conservation of energy' or the first law of thermodynamics: it says that if you put energy into a gas, the energy goes somewhere, it doesn't just disappear.

2.3.8 Heat capacity at constant pressure, enthalpy

Using the ideal gas law in the form $p\alpha = RT$, we have

$$p\frac{\mathrm{d}T}{\mathrm{d}t} = R\frac{\mathrm{d}T}{\mathrm{d}t} - \alpha\frac{\mathrm{d}p}{\mathrm{d}t} \tag{2.73}$$

which we can rewrite (2.72) as

$$c_\mathrm{p}\frac{\mathrm{d}T}{\mathrm{d}t} - \alpha\frac{\mathrm{d}p}{\mathrm{d}t} = J, \tag{2.74}$$

where

$$c_\mathrm{p} = c_\mathrm{v} + R = \frac{s+2}{2}R \tag{2.75}$$

is the *specific heat capacity at constant pressure* and s is the effective number of degrees of freedom per molecule. If pressure is constant, then heating the gas raises the temperature at a rate slower than if volume is constant; the extra energy is lost to the environment by expansion.

The *enthalpy* per unit mass is defined as

$$h = u + p\alpha = c_\mathrm{v}T + RT = c_\mathrm{p}T, \tag{2.76}$$

where u is the internal energy per unit mass. For a constant pressure system,

$$\frac{\mathrm{d}h}{\mathrm{d}t} = c_\mathrm{p}\frac{\mathrm{d}T}{\mathrm{d}t} = J. \tag{2.77}$$

Thus, enthalpy may be thought of as the amount of energy left over after the system has done expansion work to keep the pressure constant.

2.4 Entropy and the second law of thermodynamics

2.4.1 Definition of entropy

Again using the ideal gas law, (2.74) can be written

$$\frac{1}{T}\frac{\mathrm{d}T}{\mathrm{d}t} - \frac{R}{c_\mathrm{p}}\frac{1}{p}\frac{\mathrm{d}p}{\mathrm{d}t} = \frac{\mathrm{d}}{\mathrm{d}t}\ln\left(Tp^{-R/c_\mathrm{p}}\right) = \frac{J}{c_\mathrm{p}T}, \tag{2.78}$$

or equivalently

$$\frac{\mathrm{d}s}{\mathrm{d}t} = \frac{J}{T} \tag{2.79}$$

where

$$s \equiv c_\mathrm{p}\ln\left(Tp^{-R/c_\mathrm{p}}\right) + \text{const.} \tag{2.80}$$

is the *entropy* per unit mass. Choosing a fixed reference pressure p_0 and temperature T_0, and defining the constant in (2.80) to be $-c_\mathrm{p}\ln(T_0 p_0^{-R/c_\mathrm{p}})$, then

$$s = c_\mathrm{p}\ln\left(\frac{Tp^{-R/c_\mathrm{p}}}{T_0 p_0^{-R/c_\mathrm{p}}}\right). \tag{2.81}$$

Equation (2.79) says that heating is a source of entropy. From the energy point of view, heating and working are equivalent—they both change the internal energy by the same amount. But from the entropy point of view, they are quite distinct: energy exchange with the environment through disorganized molecular motion (heating) changes the entropy of the gas, while energy exchange through organized, macroscopic motion (working) leaves the entropy unchanged.

2.4.2 The second law of thermodynamics

Processes in which there is no heating ($J = 0$) are called *adiabatic*. Equation (2.79) suggests that entropy is constant in an adiabatic process. Actually, this is not really true: many strictly adiabatic processes do not conserve entropy. Here is a simple example. Consider a gas in an insulated piston-and-cylinder system as in figure 2.9. The gas is initially in equilibrium and the piston is then lifted instantaneously by some distance. The gas will freely expand to fill the container. Since the gas does no work in expansion, its internal energy (and thus temperature) does not change. However the density, and thus the pressure, have both decreased. Therefore the entropy must *increase*.

So what is wrong here? Why does (2.79) mislead us? The answer is that (2.79), which derives from (2.72), assumes that there is a well-defined and uniform temperature and pressure at each instant—it assumes that the system is progressing through a sequence of thermodynamic equilibrium states. Such a process is called *quasi-static*. Furthermore, (2.79) assumes that any volume change involves work through the term $p\mathrm{d}\alpha/\mathrm{d}t$. Both assumptions are violated during free expansion: after the piston is lifted, the gas is out of equilibrium (the empty half of the cylinder will fill with faster molecules first, so it will not have a Maxwell–Boltzmann distribution), and the gas expands but does no work. Thus, (2.79) simply does not apply to the free expansion problem.

Free expansion is an example of an *irreversible* process: the molecules will not spontaneously return to the lower half of the cylinder and, if we compress the gas back to its original volume, its temperature will be higher than originally. Some processes *are*, however, reversible: if we allow the gas to expand quasistatically by lowering the external pressure very gradually, then we can recompress it again just as slowly and return to the original state with no change in entropy.

Reversible processes can be defined to be those which leave the total entropy of the Universe[2] unchanged. Free expansion is an irreversible process leading to an increase in total entropy. Do *all* irreversible processes increase total entropy? The answer to this question (yes) is the second law of thermodynamics: *the entropy of an isolated system (such as the Universe) cannot decrease.*

Where does all this leave us as atmospheric scientists? Motion in a gas can be considered quasi-static if its speed is much slower than the speed of sound. Since this is true for all naturally occurring motion in the atmosphere, and since parcel expansion in the atmosphere is invariably connected with work done against the surrounding air, we can always safely assume that (2.79) is valid. As a corollary, any atmospheric motion which is adiabatic will entail zero entropy change and is therefore reversible. In what follows, *we will always assume that adiabatic processes are isentropic (i.e. they conserve entropy)* (but remember that this is an approximation, not a law of nature). As it turns out, many atmospheric motions of importance—such as motion associated with synoptic-scale mid-latitude cyclones—are fast enough to be considered adiabatic (i.e. air parcels will move substantial distances before diabatic processes such as radiation have time to act). Thus isentropic motion occupies a very important place in atmospheric science.

While adiabatic processes can generally be considered reversible, many *diabatic* processes in the atmosphere are strongly irreversible. An important example is *thermal mixing*, in which parcels at different temperatures are brought together for long enough to exchange heat until they equilibrate at a common temperature. This process leads to an irreversible gain in entropy, as required by the second law and discussed further in the following section. Another example is the mixing of a dry and a moist parcel at the same temperature; in this case, the process can be seen as the free expansion of the water vapour from the moist parcel into the dry one, and again entails an increase in entropy. How important this irreversibility is to the overall functioning of the atmosphere remains a matter of debate; we return to this point in section 2.4.4.

[2] The 'Universe' can be thought of as a large isolated system surrounding the particular system we are focusing on. 'Isolated' means that the system cannot exchange energy with any other system, either through heating or working.

2.4.3 Thermodynamic equilibrium and heat conduction

The second law implies a ratcheting effect for entropy: once a certain entropy has been achieved, there is no going back. This gives a directionality to natural processes. A system that is evolving in time and experiencing many different states will gradually increase its entropy. It will stop evolving once it has reached a state of maximum entropy. Thus, thermodynamic equilibrium is characterized by having the highest entropy compatible with the constraints on the system.

When the constraints are weak, thermodynamic equilibrium is isothermal—the temperature is uniform throughout the system. To see this, consider a closed system consisting of two parts, 1 and 2, with temperatures T_1 and T_2, respectively, which exchange heat at rate Q. Say that heat is going from 1 to 2. In time $\mathrm{d}t$, the entropy changes will be $\mathrm{d}s_1 = -Q\mathrm{d}t/T_1$ and $\mathrm{d}s_2 = Q\mathrm{d}t/T_2$. Since entropy must increase, $\mathrm{d}s_1 + \mathrm{d}s_2 > 0$ which implies $T_1 > T_2$. This has two consequences: (1) heat can only flow from the warmer to the colder body and (2) heat flow will stop when $T_1 = T_2$, at which point the system is isothermal and can gain no more entropy.

But heat conduction within a gas is controlled by molecular motions at the microscopic level. Are we sure that these motions will behave in the proper way to satisfy the second law? A rough estimate for the molecular heat flux in an ideal gas serves to illustrate that the answer is yes. Consider again a gas in a pipe, as in figure 2.3. The flux of KE in any direction within the pipe is

$$\sim n\langle v\rangle \frac{1}{2} m\langle v^2\rangle \sim \langle v\rangle nkT \sim p\sqrt{T} \tag{2.82}$$

where n is number density and p is pressure, and we have used the ideal gas law and the relation $\langle v\rangle \sim \sqrt{T}$. Now consider specifically the flux from left to right across surface S, which is due mostly to molecules coming from a region of width λ (the mean free path) to the left of S; since such molecules generally do not collide, they will conserve their energy. The flux is roughly

$$f_+ \sim p(x-\lambda)\sqrt{T(x-\lambda)} \sim p\sqrt{T} - \frac{p}{2\sqrt{T}}\frac{\mathrm{d}T}{\mathrm{d}x}\lambda \tag{2.83}$$

where x is the position of the surface S. To obtain the last expression, we have expanded to first order and assumed the gas is not accelerating, so that $\mathrm{d}p/\mathrm{d}x = 0$. With an analogous estimate for the flux f_- moving from right to left across S we obtain the *net* rightward flux

$$f = f_+ - f_- \sim -\frac{p}{\sqrt{T}}\frac{\mathrm{d}T}{\mathrm{d}x}\lambda \sim -n\langle v\rangle\lambda\frac{\mathrm{d}T}{\mathrm{d}x}. \tag{2.84}$$

Thus, heat flows *down* the temperature gradient (from hot to cold) and ceases to flow when temperature is uniform, exactly as required by the second law. A more precise calculation using the full apparatus of kinetic theory gives the same qualitative result.

Exercise 2.4.3: Extend the argument above to show that (2.84) also applies to a vertical column of air in hydrostatic equilibrium.

2.4.4 The Carnot engine

A *heat engine* is a mechanical device which takes thermal energy from an outside source and tries to convert as much of it as possible into useful work. The *efficiency* of the engine may be defined as the ratio of heat intake to work production. Understanding the limitations on efficiency is obviously of great interest.

The simplest way to turn heating into working is to heat a gas and allow it to expand. In any real machine which needs to operate continuously, the gas cannot be allowed to expand forever, so it must be periodically recompressed. Thus, heat engines work in an expansion–compression cycle. The *Carnot engine* is an idealized heat engine which allows particularly clear theoretical insight into the question of efficiency. The cycle of a Carnot engine can be illustrated using a p–α diagram (figure 2.10). Beginning at point 1, the cycle consists of an isothermal expansion at temperature T_{hot}, followed by adiabatic expansion which lowers the temperature to T_{cold}. The cycle is closed by an isothermal compression at T_{cold} and an adiabatic compression bringing the temperature back to T_{hot}.

If we integrate the first law,

$$c_{\text{v}}\frac{\mathrm{d}T}{\mathrm{d}t} + W = J \tag{2.85}$$

over a cycle, we obtain

$$\Delta W = \Delta J = \Delta J_{\text{hot}} - \Delta J_{\text{cold}} \tag{2.86}$$

where $\Delta W = \int_{\text{cycle}} W\mathrm{d}t$ is the total work done, ΔJ is the total heat entering the system over the cycle, ΔJ_{hot} is the heat entering the system during isothermal expansion, and ΔJ_{cold} is the heat *exiting* the system during isothermal compression

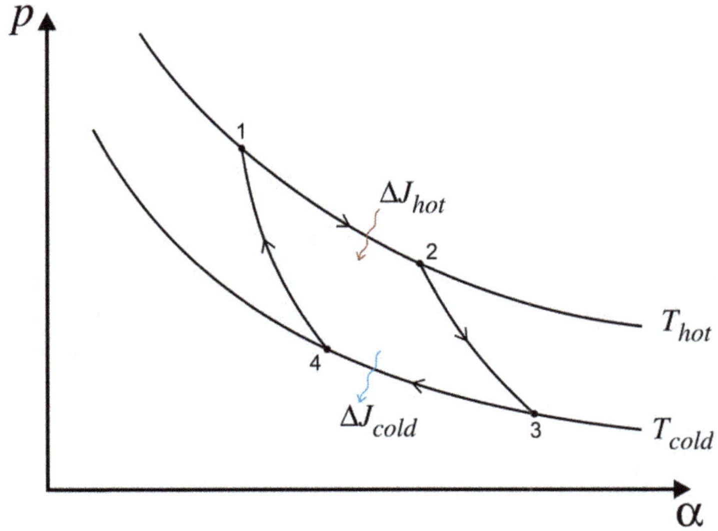

Figure 2.10. The Carnot engine.

(no heat enters during the adiabatic segments, obviously). Since ΔJ_{hot} is the engine's heat intake in each cycle, we can define the efficiency as

$$\eta = \frac{\Delta W}{\Delta J_{\text{hot}}} = 1 - \frac{\Delta J_{\text{cold}}}{\Delta J_{\text{hot}}}. \tag{2.87}$$

The initial and final temperature and pressure of the gas are the same, so the total entropy change of the gas over a cycle is zero. Assuming the processes to be reversible, we can write the entropy change as

$$\Delta s = \int_1^2 \frac{J\mathrm{d}t}{T_{\text{hot}}} - \int_3^4 \frac{J\mathrm{d}t}{T_{\text{cold}}} = \frac{\Delta J_{\text{hot}}}{T_{\text{hot}}} - \frac{\Delta J_{\text{cold}}}{T_{\text{cold}}}. \tag{2.88}$$

Now comes the key point: entropy is a *state variable*, meaning that it is purely a function of thermodynamic state, identified by p, T and ρ. At the end of the cycle, the system returns to its initial state and all thermodynamic quantities return to their initial values. Therefore, there is *no* overall change in the systems' entropy over the cycle:

$$\Delta s = 0 \tag{2.89}$$

which implies

$$\frac{\Delta J_{\text{cold}}}{\Delta J_{\text{hot}}} = \frac{T_{\text{cold}}}{T_{\text{hot}}}, \tag{2.90}$$

so finally:

$$\eta = 1 - \frac{T_{\text{cold}}}{T_{\text{hot}}}, \tag{2.91}$$

which is known as the *Carnot efficiency.*

Up to now, the second law has played no role. This is because we have considered only reversible processes, which are an idealized abstraction. In reality, all processes entail some amount of irreversibility, and the second law states that they must therefore generate some entropy. Thus, we can write the entropy integral over the cycle as

$$\frac{\Delta J_{\text{hot}}}{T_{\text{hot}}} - \frac{\Delta J_{\text{cold}}}{T_{\text{cold}}} + \Delta s_{\text{irr}} = 0 \tag{2.92}$$

where Δs_{irr} is the (positive) entropy generated by irreversibility. Note that the total must still equal 0, since entropy is still a state variable: what happens is that the extra entropy generated by irreversibility is ejected during the cold compression phase, requiring ΔJ_{cold} to be larger than before. Using (2.86) to eliminate ΔJ_{cold}, we find

$$\frac{\Delta W}{\Delta J_{\text{hot}}} = \eta - \frac{T_{\text{cold}}}{\Delta J_{\text{hot}}} \Delta s_{\text{irr}}. \tag{2.93}$$

Since the second law requires the second term on the rhs to be negative, *the Carnot efficiency is the maximum possible efficiency* for any heat engine operating between two specified temperatures and is achieved only if the engine is fully reversible.

The atmosphere can be treated as a heat engine: energy is deposited by solar radiation at the relatively warm surface and emitted back out to space from higher, colder altitudes (see chapter 5); part of this energy flux is used to perform the work of driving atmospheric motion against friction. If irreversibility can be neglected, then knowledge of the input and output temperatures and heat flux can be used to predict the KE of the entire atmosphere or of important sub-systems like individual cyclones in a simple way, bypassing the need to actually solve the equations of motion. There is an animated debate regarding the validity of this approximation, however, with some authors claiming that irreversibility reduces the thermodynamic efficiency of the atmosphere by an order of magnitude.

2.4.5 Potential temperature

Meteorologists generally do not work directly with entropy. Instead, we define the *potential temperature*

$$\Theta = T\left(\frac{p}{p_0}\right)^{-R/c_p}, \tag{2.94}$$

where p_0 is a fixed reference pressure. It is related to entropy by

$$\Theta = T_0\, e^{s/c_p} \tag{2.95}$$

where T_0 is the reference temperature. Just like entropy, potential temperature is conserved under adiabatic, reversible transformations. It has many advantages in atmospheric applications, mostly related to its simple physical interpretation: it is the temperature a parcel of air will have if brought adiabatically to pressure p_0. To see this, note that for an adiabatic process,

$$T_0 p_0^{-R/c_p} = T p^{-R/c_p} \tag{2.96}$$

where T and p are the starting temperature and pressure, and T_0 is the temperature once pressure reaches p_0. Then

$$T_0 = T\left(\frac{p}{p_0}\right)^{-R/c_p} = \Theta. \tag{2.97}$$

If we assume hydrostatic equilibrium, (2.74) can be written

$$c_p\frac{dT}{dt} + g\frac{dz}{dt} = J, \tag{2.98}$$

which implies that the quantity

$$\eta = c_p T + gz \tag{2.99}$$

is also conserved under reversible adiabatic motion. η is called the *dry static energy*: 'static' because there is no KE contribution and 'dry' because there is no latent heat.

2.4.6 The dry adiabat

Now consider an atmosphere whose vertical temperature structure is such that Θ is constant. Such an atmosphere has the following interesting property: any parcel brought adiabatically to the reference pressure p_0 will have the *same* temperature on arrival as the ambient air. In other words, the temperature structure is *invariant with respect to adiabatic rearrangements.* However much you stir this atmosphere, the temperature structure does not change. In an incompressible fluid (e.g. water), this is only possible if the temperature is uniform. Since the atmosphere is compressible, however, the invariant profile actually has temperature decreasing with height, at a constant rate known as the *dry adiabatic lapse rate* Γ_d. The 'dry' here really means that no condensation occurs upon parcel displacement—the parcel can contain some moisture, just not too much for it to condense.

The value of Γ_d is found by noting that for an atmosphere on the dry adiabat,

$$\frac{d\Theta}{dz} = \left(\frac{p}{p_0}\right)^{-R/c_p}\left(\frac{dT}{dz} + \frac{g}{c_p}\right) = 0, \tag{2.100}$$

which implies

$$\Gamma_d = \frac{g}{c_p}. \tag{2.101}$$

Figure 2.11 shows profiles of temperature and potential temperature for an atmosphere on the dry adiabat. The absolute temperature follows the straight line

$$T = T_0 - \Gamma_d z, \tag{2.102}$$

while the potential temperature is, of course, constant with height. For perfectly dry air, the dry adiabatic lapse rate is 9.8 K km^{-1}, which is considerable larger than the typical 6 K km^{-1} found in the midlatitude atmosphere.

2.4.7 Does an adiabatically lifted parcel follow the dry adiabat?

If we lift an air parcel adiabatically up from sea level, its potential temperature will stay fixed but its absolute temperature will decrease, since pressure decreases.

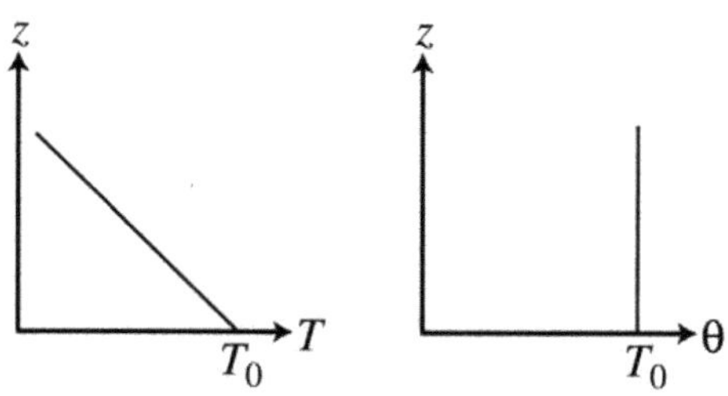

Figure 2.11. Vertical profiles of temperature (left) and potential temperature (right) in an atmosphere on the dry adiabat.

At what rate will it decrease? We can compute this by dividing the adiabatic version of (2.74) by the vertical velocity dz/dt to obtain:

$$c_p \frac{dT}{dz} - \frac{1}{\rho}\frac{dp}{dz} = 0. \tag{2.103}$$

Now assume that the parcel immediately adjusts its internal pressure to match the pressure of its surroundings, p_s—this is a very good approximation, even for parcels rising very rapidly (the equilibration is accomplished by sound waves, which travel very fast). If the surroundings are in hydrostatic equilibrium, then

$$\frac{dT}{dz} = -\frac{g}{c_p}\frac{\rho_s}{\rho} = -\frac{g}{c_p}\frac{T}{T_s}, \tag{2.104}$$

where ρ_s and T_s are the density and temperature of the surroundings. If the atmosphere is on the dry adiabat, then the parcel temperature will always match the surroundings, $T = T_s$, and so the parcel temperature will change according to the dry adiabatic lapse rate. However, the real atmosphere is generally *not* on the dry adiabat, and so the temperature inside the parcel will deviate from the dry adiabat. In practice, T generally does not differ from T_s by more than a few percent, so it is generally assumed that temperature inside an adiabatically lifted parcel *does* follow the adiabat—but you should remember that this is an approximation.

Exercise 2.4.7: Consider an atmosphere with surface temperature $T_0 = 300$ K and lapse rate $\Gamma = 6\ \mathrm{K\,km^{-1}}$. If a parcel is lifted adiabatically from the surface to the 10 km level, compute the error made in assuming that the parcel temperature follows the dry adiabat.

2.5 Static stability

2.5.1 The concept of static stability

We all know that if you put dense fluid over a lighter fluid, the denser fluid will fall down and the lighter fluid will rise to the top. In the troposphere, cold air lies over warmer air (figure 2.12). We generally think that warmer air will be less dense than colder air, so isn't the profile shown in figure 2.12 unstable? Why doesn't the warm air rise, the cold air fall and the temperature gradient reverse? The answer, of course, is that the atmosphere is compressible, and density is affected not only by temperature but also by pressure. This is different from the case, say, of water heated in a pot: pressure is irrelevant and warmer water will always be lighter than colder water.

To assess the stability of a given atmospheric temperature profile, we need to know how the density of an air parcel changes as it moves. In particular, consider the parcel marked A in figure 2.12, and imagine that some external force gives it a small upwards displacement. From the ideal gas law, we know that $\rho = p/RT$. We can safely assume that pressure inside the parcel will quickly equilibrate with the

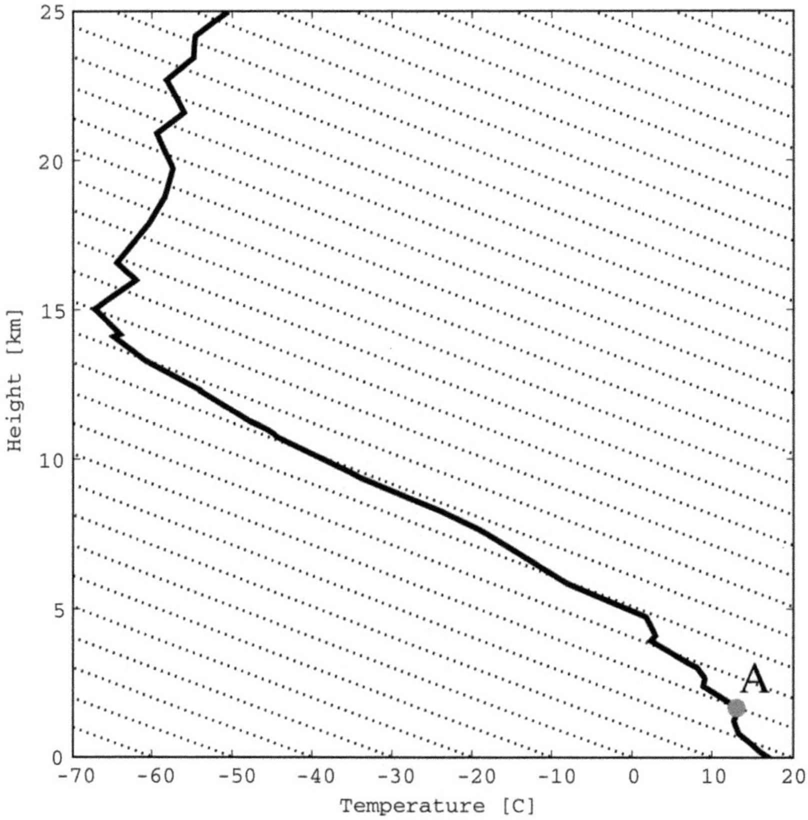

Figure 2.12. Temperature sounding at Valentia (solid line) with dry adiabats superimposed.

pressure of the surroundings. But what about temperature? If there is strong heat exchange between the parcel and its surroundings, we might expect that parcel temperature will also match the surroundings. In this case, the parcel would always equal that of the surroundings and there would be no tendency to rise or fall. But consider the opposite extreme, where there is *no* heat exchange with the surroundings. Then the parcel's temperature will approximately follow the dry adiabatic profile, shown by the dotted line passing through A. In this case, the parcel will always be *colder* and therefore denser than the surroundings, and will tend to fall back down. Thus the profile is stable: an initial upward displacement will be followed by the parcel spontaneously returning to its initial position. While there will always be *some* heat exchange between the parcel and its surroundings, it is generally a good approximation for the atmosphere to consider displacements to be adiabatic. We will formalize these ideas below, beginning with the concept of buoyancy.

2.5.2 Buoyancy and the Brunt–Väisälä frequency

Consider a fluid at rest and imagine selecting an irregularly shaped sub-volume V within the fluid. The net force on this sub-volume must be zero, otherwise it would not be stationary. The net vertical force acting on the sub-volume is the sum of the gravitational and pressure forces:

$$F_z = \int_S p \cos\theta \mathrm{d}S - g \int_V \rho \mathrm{d}V = 0 \tag{2.105}$$

where S indicates the surface of the sub-volume and θ is the angle between the normal to the surface and the vertical. Now imagine replacing this sub-volume by a foreign body (for instance, that of Archimedes) which happens to have exactly the same shape as the sub-volume of fluid that is being displaced, but a different density ρ_A. The net force is now different from zero:

$$F_z = \int_S p \cos\theta \mathrm{d}S - g \int_V \rho_A \mathrm{d}V = g \int_V (\rho - \rho_A) \mathrm{d}V. \tag{2.106}$$

Thus, if Archimedes' density is greater than that of the fluid, $F_z < 0$ and he will sink, while if his density is lesser he will float. The net upwards force experienced by Archimedes is called the *buoyancy force*.

Now, given an atmosphere with a specified temperature profile, take a parcel of air and move it adiabatically upwards from z_0 to $z = z_0 + \Delta z$ (figure 2.13). Newton's second law applied to the displaced parcel is

$$g\left(\rho_s - \rho_p\right)V = \rho_p V \frac{\mathrm{d}^2}{\mathrm{d}t^2}\Delta z. \tag{2.107}$$

The term on the left-hand side is the buoyancy force (we're assuming the parcel is small enough that its density can be considered homogeneous) and the right-hand side is mass times acceleration. Subscript 'p' refers to the parcel, while subscript 's' refers to the surroundings (or sounding). We can re-write this as

$$\frac{\mathrm{d}^2}{\mathrm{d}t^2}\Delta z = g\left(\frac{\rho_s}{\rho_p} - 1\right). \tag{2.108}$$

Since the pressure inside the parcel is always equal to the surroundings, we have

$$\frac{\rho_s}{\rho_p} = \frac{T_p}{T_s} = \frac{\Theta_p}{\Theta_s}. \tag{2.109}$$

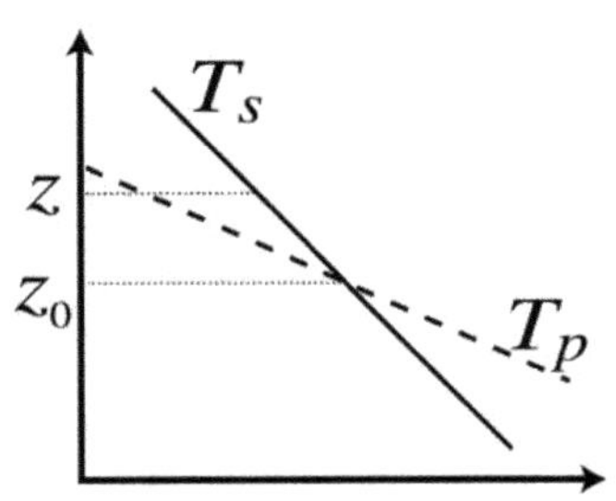

Figure 2.13. A parcel lifted adiabatically from level z_0 to level z. T_s is the temperature of the surroundings, while T_p is the temperature of the parcel and follows the dry adiabat.

Since the parcel is displaced adiabatically, Θ_p is constant and equal to its value before being displaced:

$$\Theta_p = \Theta_s(z_0). \tag{2.110}$$

The potential temperature of the surroundings, on the other hand, generally changes with height. If the displacement is small, we can describe the vertical variation using a first-order Taylor expansion:

$$\Theta_s(z) \simeq \Theta_s(z_0) + \left.\frac{d\Theta_s}{dz}\right|_{z_0} \Delta z. \tag{2.111}$$

Using a further Taylor expansion, we have

$$\frac{\Theta_p(z)}{\Theta_s(z)} \simeq \frac{\Theta_p(z)}{\Theta_s(z_0)}\left(1 - \frac{1}{\Theta_s(z_0)} \left.\frac{d\Theta_s}{dz}\right|_{z_0} \Delta z\right) = 1 - \frac{1}{\Theta_s(z_0)} \left.\frac{d\Theta_s}{dz}\right|_{z_0} \Delta z \tag{2.112}$$

where the second equality follows because the parcel is displaced adiabatically and therefore its potential temperature is constant. Finally, substituting back into (2.108) gives

$$\frac{d^2}{dt^2}\Delta z + N^2 \Delta z = 0, \tag{2.113}$$

where

$$N = \left(\frac{g}{\Theta}\frac{d\Theta}{dz}\right)^{1/2} \tag{2.114}$$

is known as the *Brunt–Väisälä frequency*. Equation (2.113) describes a harmonic oscillator with the general solution

$$\Delta z = a e^{iNt} + b e^{-iNt}, \tag{2.115}$$

where a and b are constants determined by the initial conditions. The behaviour of the solutions depends on the sign of $d\Theta/dz$:

- $d\Theta/dz > 0$ implies that N is a real number, $N = |N|$, and the solution can be written

  $$\Delta z = A \cos(|N| t + \phi), \tag{2.116}$$

  where amplitude A and phase ϕ are functions of a and b. The solution is oscillatory: after being displaced, the parcel falls back down, overshoots its initial position until positive buoyancy makes it stop and come back up, and so on. This is exactly the same behaviour as that of a ball rolling around at the bottom of a well. The system is *stable*, as gravity acts to counteract the initial perturbation.

- $d\Theta/dz < 0$ implies that N is imaginary, $N = i|N|$, and the solution takes the form

$$\Delta z = a e^{-|N|t} + b e^{|N|t}. \tag{2.117}$$

The solution is the sum of two exponentials; one damps to zero, but the other one grows indefinitely. The system is *unstable*, since gravity acts to amplify the initial perturbation. This is the same behaviour as a ball initially balanced on a hill top, which, if given a small push, will roll away never to come back.

- Finally, $d\Theta/dz = 0$ implies

$$\Delta z = a + b, \tag{2.118}$$

that is, the parcel simply remains in its perturbed position; the profile is *neutral.*

To summarize:

- If potential temperature *increases* with height, the profile is *stable.*
- If potential temperature *is constant* with height, the profile is *neutral.*
- If potential temperature *decreases* with height, the profile is *unstable.*

Thus, plotting the potential temperature shows at a glance whether a sounding is stable. Figure 2.14 shows that the sounding at Valentia is stable in the troposphere and even more stable in the stratosphere.

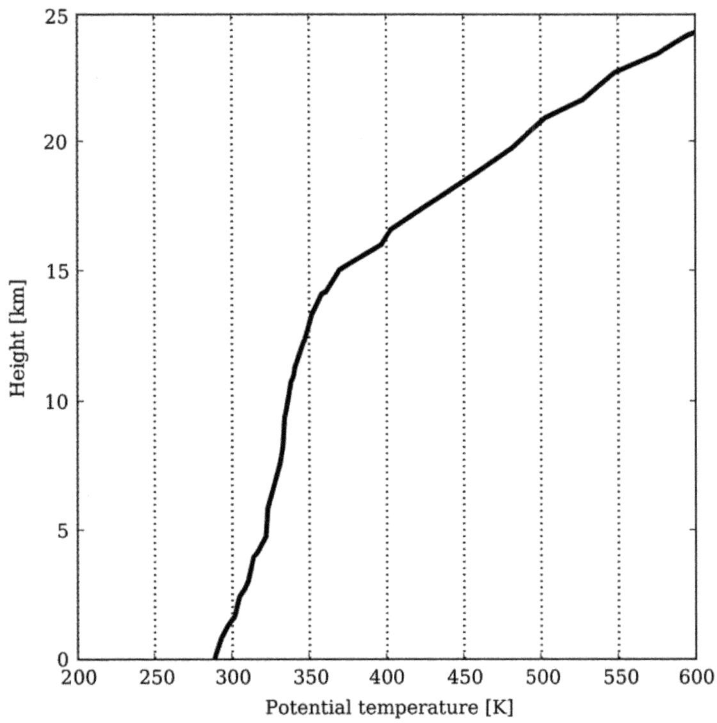

Figure 2.14. Potential temperature for the same sounding as in figure 2.12.

2.5.3 The hydrostatic approximation, continued

We are finally in a position to complete our analysis—begun in section 2.2.2—of the conditions under which the hydrostatic approximation is accurate. Suppose that the perturbation fields p' and ρ' defined in section 2.2.2 derive from an adiabatic deformation of the basic state p_0, ρ_0. Then (2.39) can be written

$$\frac{d^2 z}{dt^2} \ll N^2 \Delta z, \qquad (2.119)$$

so that using the same estimate of the acceleration as in section 2.2.2 and assuming $\Delta z \sim H$ the condition becomes

$$\left(\frac{H}{L}\right)^2 \left(\frac{U}{NH}\right)^2 \ll 1. \qquad (2.120)$$

A typical value of the Brunt–Väisälä frequency in the atmosphere is $\sim 10^{-2}\ \mathrm{s}^{-1}$ (a period of about 2 min). In frontal zones where $H \sim L \sim 1$ km, the left-hand side is ~ 1 and hydrostatic balance is therefore *not* a good approximation for the study of dynamics in such regions (though as noted in section 2.2.2 it is adequate for purely thermodynamic purposes).

2.5.4 Convective available PE

Now consider the idealized temperature profile shown in figure 2.15. This profile is unstable between points A and B, and stable everywhere else. It would be nice to have a single number which characterized the *strength* of the instability, which will be connected with the energy of the ensuing motion. To do this, consider the fate of a

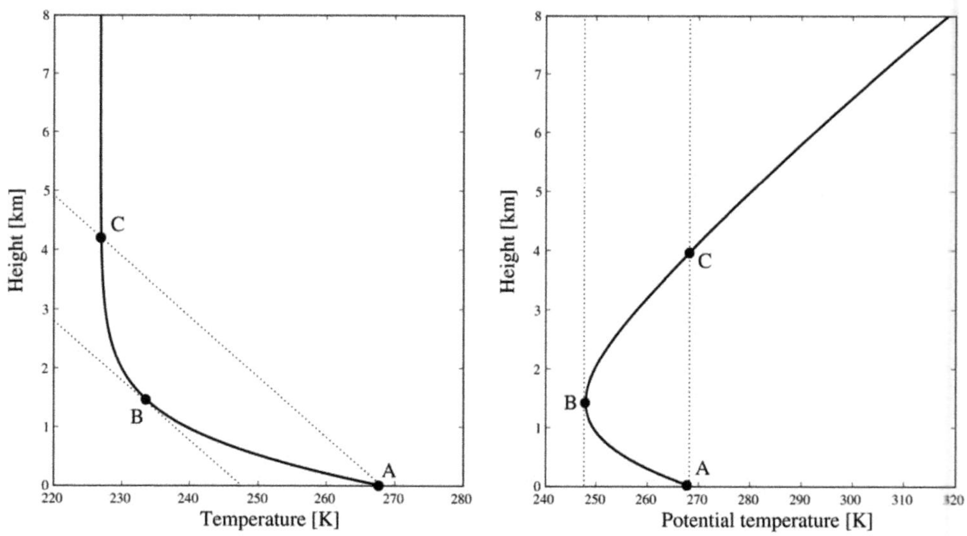

Figure 2.15. An unstable temperature profile (solid line) with selected dry adiabats (dotted).

parcel which begins at level A and is displaced upwards. The parcel will be positively buoyant all the way up to level C. This means the atmosphere will do work on the parcel between A and C. This work shows up as increased KE of the parcel: it is exactly as if the parcel 'fell' from A to C. This amount of energy is called *convective available PE* or CAPE. A parcel starting from some point between A and B will be positively buoyant up to a corresponding point between B and C and will also gain KE, though the energy will be less since the parcel will 'fall' a shorter distance. Thus, the energy gained by the parcel travelling from A to C is a maximum and provides a reasonable measure of the overall instability of the profile.

To compute CAPE, we need to compute the work done by the buoyancy force (2.106) between A and C:

$$\int_{z\mathrm{A}}^{z\mathrm{C}} g\left(\frac{\rho_\mathrm{s}}{\rho_\mathrm{p}} - 1\right)\mathrm{d}z = \int_{z\mathrm{A}}^{z\mathrm{C}} (\alpha_\mathrm{p} - \alpha_\mathrm{s})g\rho_\mathrm{s}\mathrm{d}z, \qquad (2.121)$$

which using hydrostatic equilibrium and the ideal gas law, gives

$$\mathrm{CAPE} = R\int_{p\mathrm{C}}^{p\mathrm{A}} \left(T_\mathrm{p} - T_\mathrm{s}\right)\mathrm{d}\ln p. \qquad (2.122)$$

As noted above, the work done on the parcel will show up as KE of the parcel, so

$$v_z(z_\mathrm{C}) = \sqrt{2\,\mathrm{CAPE}}, \qquad (2.123)$$

where v_z is the vertical velocity.

Bibliography

Kalnay E *et al* 1996 The NCEP/NCAR 40-year reanalysis project *Bull. Am. Meteor. Soc.* **77** 437–72

Chapter 3

Thermodynamics of moist air

Until now we have been dealing with a single-component gas: all molecules are identical. How do things change when two or more different types of molecule are mixed together? If none of the gases can condense, then very little changes, aside from a modified gas constant (see below). However, water can condense at the temperatures typical of Earth's atmosphere. The consequent release of latent heat and the formation of liquid droplets (clouds) has a huge impact not only on the thermodynamics of the atmosphere, but also its dynamics and radiative transfer. The presence of water makes the atmosphere enormously more complex and interesting and keeps an army of atmospheric scientists in business.

3.1 Thermodynamics of moist non-condensing air

3.1.1 The ideal gas law for a mixture of gases: partial pressures

Take two identical boxes of volume V. One contains N_1 molecules of gas 1, the other N_2 molecules of gas 2. If both gases are ideal and are kept at the same temperature, the pressures inside the boxes will be

$$p_1 = \frac{N_1}{V}kT \tag{3.1}$$

and

$$p_2 = \frac{N_2}{V}kT. \tag{3.2}$$

What happens if we put all the molecules of one box into the other? Ideal gases behave as if each molecule ignored all others. Therefore, the gas already present in the box will not even realize that new gas is being put in and the gas entering the box

doi:10.1088/978-0-7503-1052-9ch3

will behave as if the box were empty. After the new gas is put in, the pressure is simply the *sum* of the pressures due to each gas:

$$p = p_1 + p_2 = \frac{N_1 + N_2}{V} kT. \tag{3.3}$$

This is known as *Dalton's law of partial pressures*: the total pressure is the sum of the pressures exerted by each of the gases if it occupied the volume alone. The law applies to a mixture with any number of components:

$$p = \frac{\sum_i N_i}{V} kT, \tag{3.4}$$

and defining the mean molecular mass

$$\langle m \rangle = \frac{\sum_i N_i m_i}{\sum_i N_i} = \sum_i f_i m_i, \tag{3.5}$$

where f_i is the number fraction of component i, we can write

$$p = \frac{\sum_i N_i m_i}{V} \frac{k}{\langle m \rangle} T = \rho R T. \tag{3.6}$$

Thus, a mixture of ideal gases behaves just like a single-component ideal gas with gas constant $R = k/\langle m \rangle$.

As we noted in section 1.1, the number density of atmospheric components other than water vapour is essentially constant in space and time. It is thus useful to define a gas constant for *dry air*

$$R_d = \frac{k}{\sum_{i=N2,O2,Ar,\ldots} f_i m_i} = \frac{k}{m_d}, \tag{3.7}$$

where the sum extends to all components other than water vapour and the gas constant for water vapour

$$R_v = \frac{k}{m_v}. \tag{3.8}$$

3.1.2 Six names for moisture

From the point of view of thermodynamics, the atmosphere may be considered a variable mixture of two components, dry air and water vapour. This mixture is called *moist air*. There are many equivalent and widely used ways of specifying the amount of moisture in dry air. We list them here for later reference, together with the typical units.

- Number density n_v (molecules m^{-3})
- Partial pressure $e = n_v kT$ (hPa)
- Number fraction, also called volume mixing ratio, $f_v = n_v/(n_v + n_d) = e/p$ (%)
- Mass density $\rho_v = n_v m_v$ (kg m^{-3})
- Specific humidity $q = \rho_v/\rho$ (g kg^{-1})
- Mass mixing ratio $w = \rho_v/\rho_d$ (g kg^{-1})

Note the subtle distinction between specific humidity (mass density of water divided by total density of water + dry air mixture) and mass mixing ratio (density of water divided by density of dry air only).

The above definitions are all equivalent and can be expressed one as a function of the other. Some useful conversions are:

$$q = \frac{\rho_v}{\rho_v + \rho_d} = \frac{w}{w + 1}, \tag{3.9}$$

$$w = \frac{\rho_v}{\rho_d} = \frac{e/R_v T}{p_d/R_d T} = \frac{e}{p - e}\frac{R_d}{R_v} = \frac{\epsilon e}{p - e}, \tag{3.10}$$

and

$$f_v = \frac{q}{\epsilon - (\epsilon - 1)q}, \tag{3.11}$$

where

$$\epsilon = \frac{m_v}{m_d} = \frac{18}{28.9} = 0.622. \tag{3.12}$$

3.1.3 Potential temperature of moist non-condensing air

The potential temperature is defined as

$$\Theta = T\left(\frac{p}{p_0}\right)^{-R/c_p}, \tag{3.13}$$

where the R and c_p are *moist* values. Let us make the dependence on humidity explicit. First,

$$R = \frac{k}{f_d m_d + f_v m_v} = \frac{R_d}{1 + (\epsilon - 1)f_v}, \tag{3.14}$$

so using (3.11) we have

$$R = R_d\left(1 + \frac{1 - \epsilon}{\epsilon}q\right) = R_d(1 + 0.608q). \tag{3.15}$$

Because of equipartition, the heat capacity of a mixture is the sum of the heat capacities of the components. The specific heat capacity is just the mass-weighted mean:

$$c_p = c_{pd}\frac{\rho_d}{\rho} + c_{pv}\frac{\rho_v}{\rho} \tag{3.16}$$

where c_{pd} and c_{pv} are the specific heat capacities of dry air and water vapour (1005 and 1952 J K^{-1} kg^{-1}, respectively). Thus

$$c_p = c_{pd}\left(1 - q + \frac{c_{pv}}{c_{pd}}q\right) = c_{pd}(1 + 0.94q), \tag{3.17}$$

and

$$\frac{R}{c_p} = \frac{R_d}{c_{pd}}\left(\frac{1 + 0.608q}{1 + 0.94q}\right) \simeq \frac{R_d}{c_{pd}}(1 - 0.33q). \tag{3.18}$$

Since q is rarely larger than about 4% in the atmosphere, the ratio R/c_p does not deviate from the dry value by more than 1%, and this difference is safely ignored. Thus, even for moist air we take

$$\Theta = T\left(\frac{p}{p_0}\right)^{-R_d/c_{pd}}. \tag{3.19}$$

3.1.4 Virtual temperature

The ideal gas law, $p = nkT$, says that pressure depends on the number of molecules but not on their mass. This is somewhat counter-intuitive, since we might expect heavier molecules to bang more strongly on the sides of the box and produce a higher pressure. The trick is that temperature is the product of mass and mean square velocity: for a given temperature, lighter molecules travel faster and produce the same momentum flux. As a result, at fixed temperature and pressure, air becomes less dense the moister it is (since $m_v < m_d$).

A quirk of meteorology is an insatiable desire to express everything in terms of a temperature: thus, entropy is expressed as a *potential* temperature, the density effect of moisture is expressed as a *virtual* temperature, and we'll encounter a few more as go along; they are all unknown outside meteorology. To define virtual temperature, we use (3.15) above to write the ideal gas law for moist air as

$$p = \rho R_d T_v \tag{3.20}$$

where the virtual temperature is defined by

$$T_v = (1 + 0.608q)T. \tag{3.21}$$

Virtual temperature can be up to 2–3% higher than ordinary temperature, which can mean a difference of several degrees (figure 3.1).

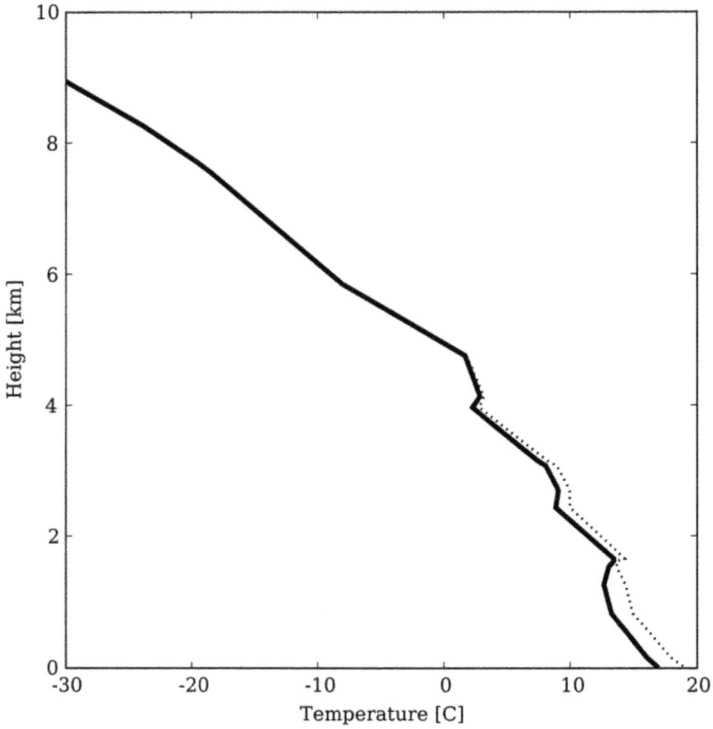

Figure 3.1. Sounding at Valentia showing temperature (solid line) and virtual temperature (dotted line).

3.1.5 Static stability of moist non-condensing air

We can take temperature madness further by defining a *virtual potential* temperature

$$\Theta_v = T_v\left(\frac{p}{p_0}\right)^{-R_d/c_{pd}}, \quad (3.22)$$

using dry values in the exponent as discussed in section 3.1.3. The beauty of virtual temperature is that we can now carry through the derivation of the Brunt–Väisälä frequency exactly as in section 2.5.2, simply replacing Θ with Θ_v everywhere (you should go through the derivation and think about why this is possible). Thus, in moist, unsaturated air the static stability criterion is

$$\frac{d\Theta_v}{dz} > 0 \quad \text{(stable)}, \qquad \frac{d\Theta_v}{dz} < 0 \quad \text{(unstable)}. \quad (3.23)$$

As can be seen by explicitly computing the vertical derivatives, this implies

$$\frac{dT_v}{dz} > -\frac{g}{c_{pd}} \quad \text{(stable)}, \qquad \frac{dT_v}{dz} < -\frac{g}{c_{pd}} \quad \text{(unstable)}. \quad (3.24)$$

Since virtual temperature is always greater than temperature, and moisture is always greater near the surface, the virtual temperature effect always acts to *destabilize* the atmosphere.

3.2 Condensation and evaporation

3.2.1 Inter-molecular forces

A key part of the definition of an ideal gas is that the component molecules do not interact with each other. This is not true in reality; the molecules of all gases can exert strong electrostatic forces on one another (figure 3.2). When two molecules are very close, they will repel each other, but when they are somewhat farther apart they will attract. The forces drop off to zero quite rapidly. If the gas is dilute, the molecules will on average be far from each other, the intermolecular forces can be neglected and the gas will behave as an ideal gas; this is the case for the atmosphere. However, if density is high enough, molecules will spend more time close to each other, intermolecular attraction will play a greater role and the gas may condense. Whether or not this happens depends on the mean separation between molecules (i.e. density) and the mean speed of the molecules (i.e. temperature): fast-moving molecules will escape each other's attraction, just as they may escape Earth's gravity (section 2.3.3). Because of its peculiar arrangement of electrons and nuclei, the water molecule has a strong, permanent electric dipole and the attractive force is very strong. For this reason, water can condense at the typical temperature and (partial) pressure at which it is found in the atmosphere, unlike other major constituents which do not have permanent dipoles.

3.2.2 Saturation vapour pressure

Consider the thought experiment illustrated in figure 3.3. We put some liquid water in a cylinder and cap it with a perfectly fitting piston, with no air between the piston and the cylinder. Now we yank the piston upwards. If the cross-section of the cylinder is small enough, you can do this with your hand. What happens next? In the first instant, there will be a vacuum between the water and the piston. But this will not last long; water molecules will soon escape from the liquid and fill the cavity with vapour. This process is very much like the escape-from-gravity process of section 2.3.3. In the liquid, the mean distance between molecules is roughly where the force curve in figure 3.2 crosses zero. Just as in the gas phase, the molecules are jiggling around—they have a mean KE, measured by the temperature of the liquid. As two

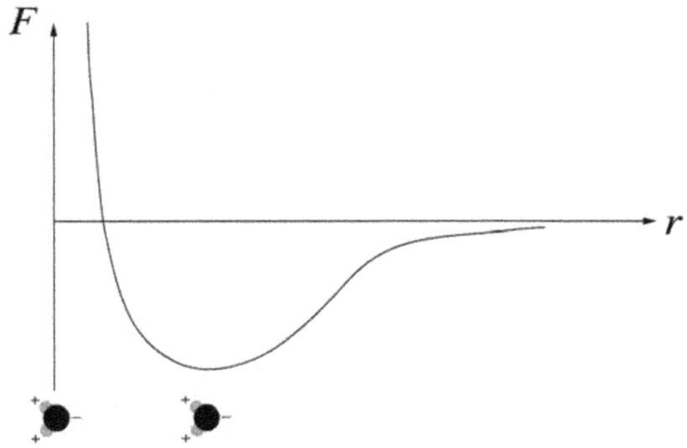

Figure 3.2. A schematic of the intermolecular force between two water molecules.

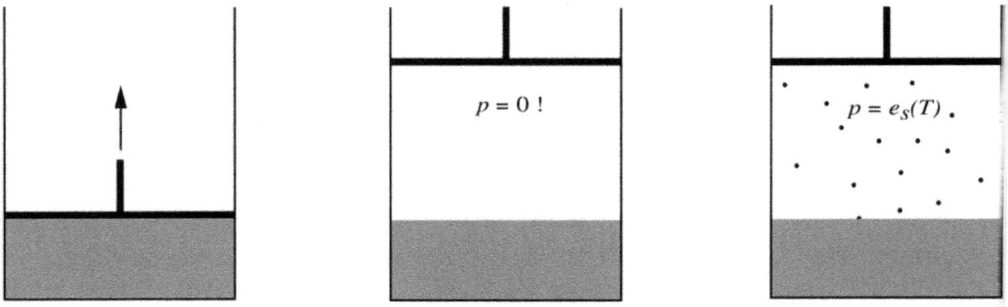

Figure 3.3. A cylinder with a tight-fitting piston containing liquid water. When the piston is lifted, water molecules leave the liquid and fill the empty space until their density is such that the flux of outgoing molecules matches the flux of molecules coming back in to the liquid.

molecules move apart, the intermolecular force does work on them, resulting in conversion of KE into electrostatic PE: the molecules slow down, and eventually move back together again. However, molecular velocities are again distributed according to the Maxwell–Boltzmann distribution, and there will be a finite number of molecules with sufficient KE to entirely overcome the attraction and escape to infinity, joining the gas phase.

As molecules fill the cavity and bounce around in it, some of them will re-enter the liquid. Eventually, a steady state will be reached where the flux of particles leaving the liquid matches the ingoing flux. The pressure in the cavity at this point is called the *saturation vapour pressure* and given the symbol e_s. The saturation vapour pressure clearly depends on temperature and will increase with increasing temperature (see section 3.2.6). It also depends on the strength of the intermolecular forces and will be greater for liquids with weaker intermolecular attraction. As we will see later, it also depends on the shape of the liquid–gas interface and on the purity of the liquid. Thus we can state that *the saturation vapour pressure of water above a plane surface of pure liquid depends only on temperature*, $e_s = e_s(T)$.

3.2.3 Relative humidity and dew-point temperature

In section 3.1.2 we gave six ways of specifying the humidity content of air. The concept of saturation vapour pressure allows for two more definitions.

- *Relative humidity.* Given the saturation vapour pressure e_s, we can use (3.10) to define the saturation mixing ratio

$$w_s = \frac{\epsilon e_s}{p - e_s}. \tag{3.25}$$

Relative humidity r is then defined as the ratio of the actual mixing ratio to the saturation value:

$$r = \frac{w}{w_s}, \tag{3.26}$$

which again using (3.10) can be written

$$r = \frac{e}{e_s}\frac{p - e_s}{p - e} \simeq \frac{e}{e_s}, \tag{3.27}$$

since e/p is never more than a few percent.

- *Dew point temperature* T_d is defined as the temperature to which an air parcel must be cooled at constant pressure to achieve saturation and is given by

$$e = e_s(T_d). \tag{3.28}$$

3.2.4 Latent heat of vapourization

It is clear from the discussion above that energy transformations play a key role in evaporation and condensation. Only the most energetic molecules can leave the liquid and so evaporation implies a net loss of KE for the liquid and a consequent lowering of the temperature; this is why sweating cools you down. In exactly the reverse process, the molecules accelerate as they enter the fluid, attaining above-average KE and hence increasing its temperature; this is the famous 'release of latent heat'.

How can we quantify the energies involved? Consider the situation in the right-most panel in figure 3.3, but now imagine that the outside pressure is $e_s(T)$ and that the cylinder is adiabatic. Now add a little heat to the liquid (how you do this through an adiabatic container is part of the magic of thought experiments): instead of raising its temperature, the heat will go into evaporating some of the liquid. The energy required to evaporate 1 kg of liquid under these conditions is called the *latent heat of vapourization* and is given by

$$\ell_v = u_v - u_l + p(\alpha_v - \alpha_l). \tag{3.29}$$

The term $u_v - u_l$ is the difference in internal energy (per unit mass) between the vapour and liquid; it is the total energy required to overcome the molecular attractions between 1 kg worth of molecules. The second term is the work done against the external pressure to expand the cavity and keep it at constant pressure: $\alpha_v - \alpha_l$ is the change in volume when 1 kg of liquid water is converted into vapour at constant pressure. Recalling that the enthalpy per unit mass is

$$h = u + p\alpha, \tag{3.30}$$

we can also write

$$\ell_v = h_v - h_l. \tag{3.31}$$

Calculating ℓ_v from first principles, assuming a given structure for intermolecular forces, is a very difficult task. Fortunately, we don't need to do it; we can just *measure* its value. It turns out to be about 2.5×10^6 J kg^{-1} (at 0 °C). This is a lot of energy; the heat capacity of liquid water is 4218 J kg^{-1} K^{-1}, so with the heat released by condensing 1 kg of water vapour you could bring more than 7 kg of water from room temperature to the boiling point.

Evaporation of water from ice—that is, the direct passage from the solid to the vapour phase—is called *sublimation*. Ice is a more stable state than liquid water; it takes more energy to extract a water molecule from ice than from liquid water and the latent heat of sublimation is correspondingly larger, measured to be $2.834 \times 10^6\,\mathrm{J\,kg^{-1}\,K^{-1}}$ at 0 °C.

3.2.5 Wet-bulb temperature

Consider a drop of rain falling in a column of *unsaturated* air with uniform temperature and humidity. Let's assume that the drop's temperature is initially the same as that of the surroundings. Since the surroundings are unsaturated, the drop cools by evaporation as it falls, but is warmed by contact with the air. Eventually, the drop will reach a steady-state temperature. What is this temperature?

To answer this, consider the situation once the drop has reached its steady-state temperature, which we will call T_w. As the drop falls into some new, undisturbed air, it will find itself out of equilibrium with the air around it. The drop's surface will quickly equilibrate with a very thin layer of air surrounding it (the thickness of this layer will be a few times the distance a molecule can travel before colliding with another molecule, typically 1 μm or less). Here, 'equilibrate' means that by exchanging energy and molecules, the drop's surface and the thin air layer quickly reach the same temperature and the air becomes saturated. However, we are *assuming* that the drop's temperature is in steady state, so after equilibration the surface of the drop will have come back to temperature T_w. Thus the air around the molecule will *also* have temperature T_w and its vapour pressure will be the saturation vapour pressure at T_w. As the drop moves on, the saturated layer is stripped away and replaced by fresh unsaturated air, and the whole process is repeated. Thus the drop continuously loses mass, leaving behind itself a trail of saturated air at T_w.

Based on this picture, we can see that T_w is *the temperature to which an air parcel drops when it gives up the energy required to evaporate just enough water to bring it to saturation.* T_w is called the *wet-bulb temperature*. To make the definition precise, we make two further assumptions: that the process occurs at constant pressure and that the water which is being evaporated is already at temperature T_w. We can now write down an equation relating wet-bulb temperature to temperature and mixing ratio. The amount of energy (more precisely, enthalpy) needed to raise the temperature of a moist parcel from T_w to T at constant pressure is

$$\left(M_d c_{pd} + M_v c_{pv}\right)(T - T_w), \tag{3.32}$$

where M_d and M_v are the masses of dry air and moisture in the parcel. The amount of enthalpy needed to evaporate enough water to make the parcel saturated is

$$\ell_v(M_{vs} - M_v) \tag{3.33}$$

where M_{vs} is the mass of water in the parcel when it is saturated at the wet-bulb temperature. Setting (3.32) equal to (3.33) and dividing by M_d gives

$$(c_{pd} + wc_{pv})(T - T_w) = \ell_v(w_s - w). \tag{3.34}$$

Neglecting the wc_{pv} contribution on the left-hand side, we have

$$w = w_{\mathrm{s}}(T_{\mathrm{w}}) - \frac{c_{\mathrm{pd}}}{\ell_{\mathrm{v}}}(T - T_{\mathrm{w}}), \tag{3.35}$$

or using $w \simeq \epsilon e/p$,

$$e = e_{\mathrm{s}}(T_{\mathrm{w}}) - \frac{pc_{\mathrm{pd}}}{\epsilon\ell_{\mathrm{v}}}(T - T_{\mathrm{w}}) \tag{3.36}$$

which is known as the *psychrometric equation.*

If we have a good way of measuring T_{w}, we have a way to determine the humidity of air ('psychro' is Greek for 'cold', so 'psychrometric' means 'measuring the cold'). Humidity is tricky to measure directly, but measuring the temperature is easy. In a *sling psychrometer*, the bulb of a thermometer is wrapped in gauze soaked in water and then spun around; this approximates the situation for the falling raindrop and the steady-state temperature reading approximates T_{w} for the ambient air. From a forecasting point of view, if T_{w} falls below zero in a layer near the surface, then snow or hail falling through the layer will reach the surface without melting, while rain falling through this layer may reach the surface as freezing rain—supercooled drops that freeze on impact.

3.2.6 The Clausius–Clapeyron equation

We saw above that the saturation vapour pressure above a flat surface of pure liquid depends only on temperature. The Clausius–Clapeyron equation provides an explicit expression for the rate of change of saturation vapour pressure with temperature. It can be derived through a simple application of the Carnot engine, discussed in section 2.4.4. In this case, the working fluid is water vapour in equilibrium with liquid water, as in the rightmost panel of figure 3.3 (note that there is no air in the cylinder, only water in liquid and vapour forms).

The operating cycle in pressure–volume space is shown in figure 3.4. In the initial state (point 1) the system is in equilibrium at temperature T and vapour pressure e_{s}. We now heat the liquid–vapour mixture diabatically (and reversibly); all the energy thus provided goes into evaporating water—i.e. extracting molecules from the liquid phase—with no change in temperature and since the temperature is not changing neither does the equilibrium vapour pressure. There is however a change in volume, since the vapour occupies more room than the liquid, so overall the system expands from α_1 to $\alpha_2 > \alpha_1$. Next, we allow the system to expand adiabatically by a small amount, so that the temperature and pressure drop by infinitesimal amounts $\mathrm{d}T$ and $\mathrm{d}e_{\mathrm{s}}$, respectively. We then cool the system diabatically and finally compress it adiabatically to return to the initial state.

Now suppose that the total energy provided during the heating stage, ΔJ_{hot}, is just enough to evaporate 1 kg of water: then by definition $\Delta J_{\mathrm{hot}} = \ell_{\mathrm{v}}$, the latent heat of vapourization, and the change in specific volume is $\alpha_2 - \alpha_1 = \alpha_{\mathrm{v}} - \alpha_\ell$ (see section 3.2.4). Furthermore, the total work performed—which is contributed only by the isothermal stages, since the adiabatic stages are infinitesimal—is

$$\Delta W = (\alpha_{\mathrm{v}} - \alpha_\ell)\, \mathrm{d}e_{\mathrm{s}}.$$

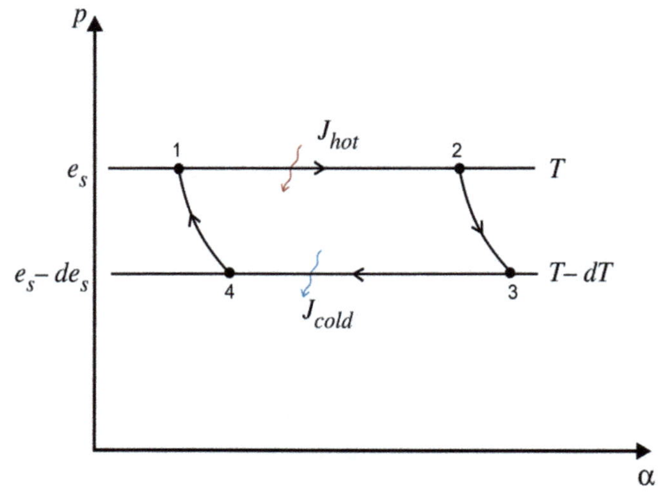

Figure 3.4. A Carnot engine used to derive the Clausius–Clapeyron equation.

We now apply the Carnot theorem in the form

$$\Delta W = \eta \Delta J_{\mathrm{hot}}$$

with thermodynamic efficiency

$$\eta = \frac{\mathrm{d}T}{T}$$

to obtain the *Clausius–Clapeyron equation*

$$\frac{\mathrm{d}e_{\mathrm{s}}}{\mathrm{d}T} = \frac{1}{T}\frac{\ell_{\mathrm{v}}}{\alpha_{\mathrm{v}} - \alpha_l}. \tag{3.37}$$

Note that the term on the right-hand side is always positive, so saturation vapour pressure always increases with temperature. Note also that an identical argument could be made using a vapour–ice mixture instead of a vapour–liquid one, so the Clausius–Clapeyron equation for ice is identical to (3.15) but using the latent heat of sublimation and the specific volume for ice.

The specific volume of liquid water (or ice) is always much smaller than that of the vapour, so we can approximate

$$\frac{\mathrm{d}e_{\mathrm{s}}}{\mathrm{d}T} \simeq \frac{1}{T}\frac{\ell_{\mathrm{v}}}{\alpha_{\mathrm{v}}} = \frac{e_{\mathrm{s}}\ell_{\mathrm{v}}}{R_{\mathrm{v}}T^2}. \tag{3.38}$$

If we assume ℓ_{v} is a constant (which it isn't), this can be integrated to give

$$e_{\mathrm{s}}(T) = e_{\mathrm{s}0}\exp\left(\frac{\ell_{\mathrm{v}}}{R_{\mathrm{v}}T_0}\right)\exp\left(-\frac{\ell_{\mathrm{v}}}{R_{\mathrm{v}}T}\right), \tag{3.39}$$

where T_0 is a reference temperature and $e_{\mathrm{s}0}$ the corresponding vapour pressure, which must be empirically determined: for reference, $T_0 = 0$ °C gives $e_{\mathrm{s}0} = 6.11$ hPa.

Equation 3.39 has a nice physical interpretation: writing

$$\frac{\ell_v}{R_v T} = \frac{m_v \ell_v}{kT}, \tag{3.40}$$

we see that the exponent is the ratio of the energy required to evaporate a single molecule to the mean KE of the molecules.

For more accurate work, we need to include the temperature dependence of ℓ_v. From (3.29) we see that

$$\frac{d\ell_v}{dT} = c_{pv} - c_l, \tag{3.41}$$

the difference in specific heats of vapour and liquid (note that since the volume of liquid water changes very little with temperature, the specific heat at constant volume and at constant pressure are almost identical, so we use the single symbol c_l to indicate both). The difference between specific heats *is* to a good approximation constant with temperature (at least over the range of interest to Earth's atmosphere), so we can integrate to obtain

$$\ell_v = \ell_{v0} + (c_{pv} - c_l)(T - T_0) \tag{3.42}$$

where ℓ_{v0} is the latent heat at some reference temperature T_0. Substituting this into (3.38) and integrating:

$$\ln\frac{e_s}{e_{s0}} = \frac{\ell_{v0} + (c_l - c_{pv})T_0}{R_v}\left(\frac{1}{T_0} - \frac{1}{T}\right) - \frac{c_l - c_{pv}}{R_v}\ln\frac{T}{T_0} \tag{3.43}$$

$$= 6808\left(\frac{1}{T_0} - \frac{1}{T}\right) - 5.09\ln\frac{T}{T_0}, \tag{3.44}$$

where the second equality is obtained by taking $T_0 = 0$ °C.

Figure 3.5 shows plots of the saturation vapour pressure over water and ice over Earth's atmospheric temperature range. Note that air in the upper troposphere, where temperatures drop below −30 °C, is essentially dry. Note also the lower saturation vapour pressure over ice than over water at the same temperature.

3.2.7 Scale height of water vapour

Since temperature decreases with height in the troposphere, e_s will also decrease with height and we expect the atmosphere to become drier with height. We can estimate the rate of decrease by assuming an atmosphere with a dry-adiabatic lapse rate, so that using (3.39):

$$e_s(z) = e_{s0}\exp\left(\frac{\ell_v}{R_v T_0}\right)\exp\left(-\frac{\ell_v}{R_v(T_0 - \Gamma_d z)}\right) \simeq e_{s0}\exp\left(-\frac{z}{H_v}\right), \tag{3.45}$$

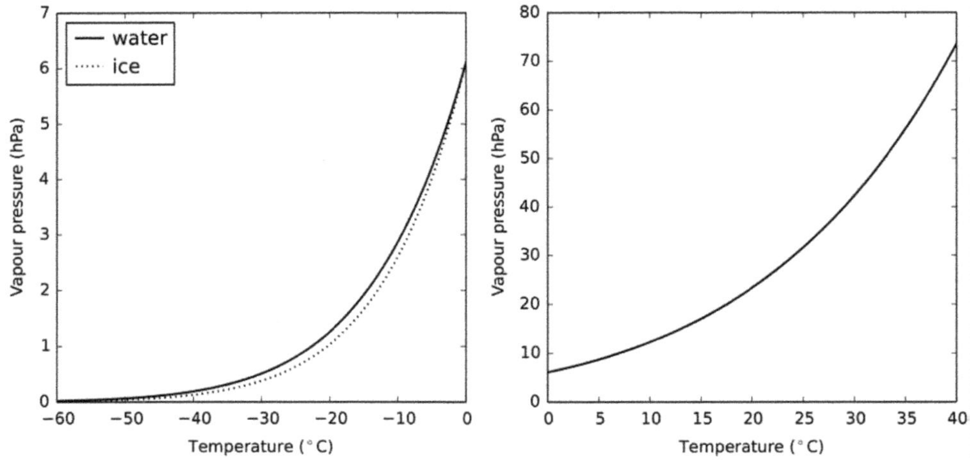

Figure 3.5. Saturation vapour pressure over water and ice over Earth's atmospheric temperature range.

where the *water vapour scale height*

$$H_v = \frac{R_v T_0^2}{\ell_v \Gamma_d} = \frac{R_v T_0}{g} \frac{c_{pd} T_0}{\ell_v} = \frac{c_{pd} T_0}{\epsilon \ell_v} H, \tag{3.46}$$

with $H = R_d T_0 / g$ the pressure scale height for dry air and ϵ given by (3.31). For Earth-like parameter values, $H_v \sim H/5 \sim 2$ km.

3.2.8 Level of cloud formation: the lifting condensation level

The most common way to form clouds on Earth is by lifting: as moist air rises, it cools and eventually becomes saturated, at which point a cloud forms. The level at which a parcel adiabatically lifted from near the surface first reaches saturation is called the *lifting condensation level* (LCL). If the parcel is lifted further, a cloud forms. On a sunny summer day, strong solar heating at the surface produces dry static instability; the consequent rising motion can produce clouds known as *fair weather cumulus*, whose sharply-defined base corresponds to the LCL.

To estimate the height of the LCL, note that the number fraction of water molecules in the parcel, f_v, remains constant as the parcel is lifted, so

$$e(z) = f_v p(z) \simeq f_v p_0 \exp(-z/H). \tag{3.47}$$

Thus the vapour pressure in the parcel decreases at a rate given by the pressure scale height. The *saturation* vapour pressure, on the other hand, decreases at the much faster rate given by the water vapour scale height. The LCL is where the two curves meet and $e = e_s$ (figure 3.6). Setting (3.47) equal to (3.45) gives

$$z_{LCL} = \frac{H H_v}{H - H_v} \ln \frac{e_{s0}}{f_v p_0} \simeq -H_v \ln r_0 \tag{3.48}$$

using $H \gg H_v$; here r_0 is the relative humidity of the parcel before lifting. If $r_0 = 0$ then $z_{LCL} = \infty$ and the parcel never saturates; if $r_0 = 1$ then $z_{LCL} = 0$ and the

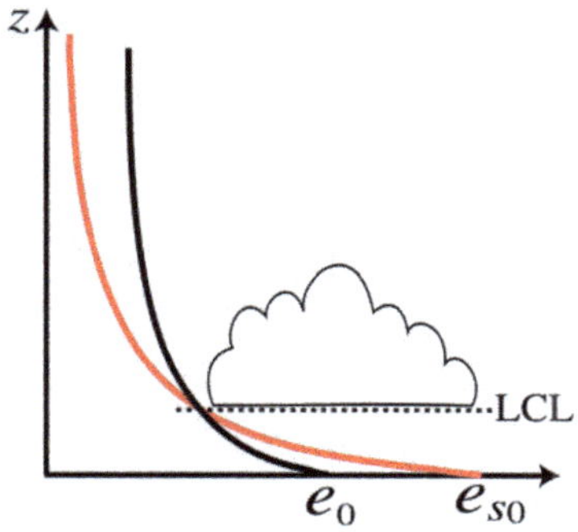

Figure 3.6. Actual vapour pressure e (black solid line) and saturation vapour pressure e_s (red line) for a parcel adiabatically lifted from the ground, where $e = e_0 = f_v\, p_0$ and $e_s = e_{s0}$. The parcel becomes cloudy when it rises above the LCL (dashed line).

parcel is already saturated at the ground. For typical near-surface r values of 70–80%, $z_{LCL} \sim 700$ m.

Note carefully that cloud formation on ascent is possible on Earth only by virtue of the fact that $H > H_v$. The opposite case, $H < H_v$, is entirely feasible; as shown by (3.46), it requires higher temperature and/or higher c_{pv}/ℓ_v, which is possible in planetary atmospheres. In this case, clouds would form on *descent*. You should look at figure 3.6 and convince yourself of this.

3.3 The moist adiabat

3.3.1 Moist entropy

In complete analogy to what we did in section 2.4.1, we will now derive an expression for the entropy of a moist air parcel in which condensation may occur. This will allow us to define a moist equivalent of the dry potential temperature. We begin with the law of conservation of energy:

$$\frac{dU}{dt} = Q - p\frac{dV}{dt}. \tag{3.49}$$

For a saturated air parcel, the internal energy can be written as

$$U = M_d u_d + M_v u_v + M_l u_l \tag{3.50}$$

where subscripts 'd', 'v' and 'l' indicate dry air, water vapour and liquid water components, respectively, and M_i is the mass of component i. Thus

$$\frac{dU}{dt} = M_d\frac{du_d}{dt} + M_v\frac{du_v}{dt} + M_l\frac{du_l}{dt} + (u_v - u_l)\frac{dM_v}{dt}, \tag{3.51}$$

since $dM_v/dt = -dM_l/dt$ (i.e. the increase of vapour mass is equal to the loss of liquid mass). Analogously, the volume V of the parcel can be written

$$V = V_l + (V - V_l) = M_l\alpha_l + M_v\alpha_v = M_l\alpha_l + M_d\alpha_d, \tag{3.52}$$

where V_l is the volume occupied by liquid droplets and $\alpha = 1/\rho = V/M$ is the specific volume; equation (3.52) essentially states that both dry air and water vapour occupy the same volume, $V - V_l$. Differentiating (3.52) and multiplying by p gives

$$p\frac{dV}{dt} = p\left(M_l\frac{d\alpha_l}{dt}\bigg|_{M_l} + M_v\frac{d\alpha_v}{dt}\bigg|_{M_v} + \alpha_v\frac{dM_v}{dt}\bigg|_{\alpha_v} + \alpha_l\frac{dM_l}{dt}\bigg|_{\alpha_l}\right) \tag{3.53}$$

$$= p\left(M_v\frac{d\alpha_v}{dt} + (\alpha_v - \alpha_l)\frac{dM_v}{dt}\right) \tag{3.54}$$

$$= p_d M_d\frac{d\alpha_d}{dt} + p_v M_v\frac{d\alpha_v}{dt} + p(\alpha_v - \alpha_l)\frac{dM_v}{dt}. \tag{3.55}$$

Equation (3.54) follows by taking

$$M_l\frac{d\alpha_l}{dt}\bigg|_{M_l} \approx 0, \tag{3.56}$$

since the volume of a fixed mass of liquid is essentially constant, and by again using $dM_v/dt = -dM_l/dt$. Equation (3.55) follows by writing p as the sum of partial pressures, $p = p_d + p_v$, and noting that

$$M_v\frac{d\alpha_v}{dt}\bigg|_{M_v} = \frac{d}{dt}(V - V_l) = M_d\frac{d\alpha_d}{dt}\bigg|_{M_d}. \tag{3.57}$$

Substituting in (3.49), we obtain

$$(M_d c_{vd} + M_v c_{vv} + M_l c_l)\frac{dT}{dt} + \ell_v\frac{dM_v}{dt} + p_d M_d\frac{d\alpha_d}{dt} + p_v M_v\frac{d\alpha_v}{dt} = Q \tag{3.58}$$

with the latent heat of vapourization ℓ_v given by (3.29).

We now write

$$p_d M_d\frac{d\alpha_d}{dt} = M_d R_d\left(\frac{dT}{dt} - \frac{T}{p_d}\frac{dp_d}{dt}\right) \tag{3.59}$$

and

$$p_v M_v\frac{d\alpha_v}{dt} = M_v R_v\left(\frac{dT}{dt} - \frac{T}{e}\frac{de}{dt}\right) \tag{3.60}$$

Substituting into (3.58) and dividing by $M_d T$ we obtain

$$(c_{pd} + wc_{pv} + w_l c_l)\frac{1}{T}\frac{dT}{dt} - \frac{R_d}{p_d}\frac{dp_d}{dt} - \frac{wR_v}{e}\frac{de}{dt} + \frac{\ell_v}{T}\frac{dw}{dt} = \frac{Q}{M_d T}, \tag{3.61}$$

where w is the vapour mixing ratio and w_l is the liquid water mixing ratio.

Now

$$\frac{\ell_v}{T}\frac{dw}{dt} = \frac{1}{T}\frac{d}{dt}(\ell_v w) - w(c_{pv} - c_l)\frac{1}{T}\frac{dT}{dt} \tag{3.62}$$

$$= \frac{d}{dt}\left(\frac{\ell_v w}{T}\right) + \frac{\ell_v w}{T^2}\frac{dT}{dt} - w(c_{pv} - c_l)\frac{1}{T}\frac{dT}{dt} \tag{3.63}$$

where we have used $d\ell_v/dT = c_{pv} - c_l$. Substituting in (3.61) and using the Clausius–Clapeyron equation gives

$$\left[c_{pd} + (w + w_l)c_l\right]\frac{1}{T}\frac{dT}{dt} - \frac{R_d}{p_d}\frac{dp_d}{dt} - \frac{wR_v}{e}\frac{de}{dt} + \frac{wR_v}{e_s}\frac{de_s}{dt} + \frac{d}{dt}\left(\frac{\ell_v w}{T}\right) = \frac{Q}{M_d T}. \tag{3.64}$$

Defining the effective heat capacity

$$c_p = c_{pd} + (w + w_l)c_l, \tag{3.65}$$

(which is a constant, since total water is conserved), we see that the *moist entropy*

$$s = c_p \ln\left[T p_d^{-R_d/c_p}\left(\frac{e}{e_s}\right)^{-wR_v/c_p} \exp\left(\frac{\ell_v w}{c_p T}\right)\right] + \text{const.} \tag{3.66}$$

is conserved under reversible adiabatic transformations. The *equivalent potential temperature*,

$$\Theta_e = T\left(\frac{p_d}{p_0}\right)^{-R_d/c_p}\left(\frac{e}{e_s}\right)^{-wR_v/c_p} \exp\left(\frac{\ell_v w}{c_p T}\right), \tag{3.67}$$

is also conserved. Physically, Θ_e is the temperature a saturated parcel would have if all the water vapour in it were to condense and the parcel were brought to sea level. Note that the above definition of Θ_e applies whether or not the parcel is saturated. However, the derivation assumes that the parcel is in thermodynamic equilibrium at all times, so *if the parcel is subsaturated there can be no liquid water*, $w_l = 0$ (any liquid water in a subsaturated parcel will evaporate irreversibly until the parcel is saturated). Note also that if the parcel is perfectly dry, then $\Theta_e = \Theta$, the dry potential temperature (note that $\lim_{x\to 0} x^x = 1$).

For a subsaturated parcel, we can also define a *saturation equivalent potential temperature* Θ_{es} as the equivalent potential temperature that an unsaturated parcel would have *if* it were saturated:

$$\Theta_{es} = T\left(\frac{p_d}{p_0}\right)^{-R_d/c_p} \exp\left(\frac{\ell_v(T)w_s(T)}{c_p T}\right), \tag{3.68}$$

with $c_p = c_{pd} + w_s(T)c_l$, where $w_s(T)$ is the saturation mixing ratio at temperature T. Note that Θ_{es} is *not* conserved as a parcel is adiabatically lifted.

3.3.2 Moist adiabatic lapse rate

In section 2.4.6 we derived the dry adiabat, defined as the vertical temperature profile in a dry atmosphere with uniform potential temperature. Here we derive the analogous result for a moist atmosphere which is *everywhere saturated.*

In a saturated atmosphere $e/e_s = 1$ and the vertical derivative of (3.67) implies

$$c_p \frac{dT}{dz} - \frac{R_d T}{p_d}\frac{dp_d}{dz} + T\frac{d}{dz}\left(\frac{\ell_v w_s}{T}\right) = 0. \tag{3.69}$$

Now let's work on the second term:

$$\frac{R_d T}{p_d}\frac{dp_d}{dz} = \frac{1}{\rho_d}\frac{d}{dz}(p - e_s) = -\frac{\rho'}{\rho_d}g - \frac{1}{\rho_d}\frac{de_s}{dT}\frac{dT}{dz} = -\frac{pR_d T}{p_d R' T'}g - \frac{\ell_v w_s}{T}\frac{dT}{dz}. \tag{3.70}$$

We have used the Clausius–Clapeyron equation in the last step; primes refer to properties of the *surroundings* (we do not use subscript 's', as in section 2.4.6, to avoid confusion with 's' for 'saturated' as used here), which are assumed to be hydrostatic. As for the dry case, we define the moist adiabat as the temperature profile an atmosphere needs to have so that temperature within an adiabatically lifted saturated parcel always matches the surroundings, $T = T'$. This still leaves an annoying $pR_d/p_d R'$ in (3.70) which we will simply approximate as 1, since the moisture mixing ratio never exceeds a few percent. With this approximation, (3.69) and (3.70) give

$$\frac{dT}{dz} = -\frac{g}{c_p} - \frac{1}{c_p}\frac{d}{dz}(\ell_v w_s). \tag{3.71}$$

Thus the moist-adiabatic lapse rate is simply the dry-adiabatic lapse rate plus a contribution due to condensation. Note that since

$$w_s = \frac{\epsilon e_s}{p - e_s} \simeq \frac{\epsilon e_s}{p} \sim \frac{\epsilon e_{s0}}{p_0}\exp\left(-\frac{z}{H_v} + \frac{z}{H}\right) \tag{3.72}$$

and $H_v < H$ (see section 3.2.7), w_s decreases exponentially with height, so the condensation term in (3.71) is positive. This means that *the moist adiabatic lapse rate is always less than the dry adiabatic.* To see this more explicitly, we can use the approximation in (3.72) to write

$$\frac{1}{w_s}\frac{dw_s}{dz} = -\frac{1}{p}\frac{dp}{dz} + \frac{1}{e_s}\frac{de_s}{dz} = \frac{g}{RT} + \frac{\ell_v}{R_v T^2}\frac{dT}{dz}. \tag{3.73}$$

Taking ℓ_v as a constant in (3.71) and using (3.73) finally gives

$$\frac{dT}{dz} = -\frac{g}{c_{pd} + (w_s + w_l)c_l}\,\frac{1 + \ell_v w_s/RT}{1 + \ell_v^2 w_s/c_{pd} R_v T^2}. \tag{3.74}$$

Thus the moist adiabatic lapse rate will be less than the dry adiabatic if $\ell_v w_s/RT < \ell_v^2 w_s/c_{pd} R_v T^2$, which implies $c_{pd} T/\epsilon\ell_v < 1$: as we saw in section 3.2.7, this is true in our atmosphere.

3.3.3 Moist adiabats and pseudoadiabats

To compute the structure of the moist adiabat explicitly, we need to integrate (3.74) in the vertical. Because of the complicated dependence of w_s on T, this needs to be done numerically. In practice, we use a finite-difference approximation to write

$$T(z + \Delta z) = T(z) - \left(\frac{g}{c_{pd} + (w_s + w_l)c_l} \frac{1 + \ell_v w_s/RT}{1 + \ell_v^2 w_s/c_{pd} R_v T^2} \right)\Bigg|_z \Delta z, \quad (3.75)$$

which allows the profile to be built up step by step given initial values of temperature and humidity. To actually perform the computation, we need to express w_s, ℓ_v and w_l as functions of z:

- For ℓ_v, we use (3.42) to write $\ell_v(z) = \ell_v(T(z))$.
- For w_s we take

$$w_s(z) = \frac{\epsilon\, e_s(z)}{p(z) - e_s(z)} \quad (3.76)$$

where $e_s(z) = e_s(T(z))$. $p(z)$ is computed using the hydrostatic equation, which consistently with the approximation $\rho' \simeq \rho_d$ made to derive (3.71) can be written

$$\frac{\mathrm{d} \ln p}{\mathrm{d}z} = -\frac{g}{R_d T} \quad (3.77)$$

and integrated numerically as above.

- For w_l (the mass mixing ratio of liquid water), the situation is more complicated. If none of the condensed water precipitates out of the parcel during ascent, then the total water content $w_t = w_s + w_l$ is constant, in which case we simply use the initial value of w_t in (3.75). This is called a *true adiabat* or *reversible adiabat*, in which entropy (and equivalent potential temperature) is exactly conserved. In realistic situations, some of the condensed water may fall out as precipitation. Exactly how much falls out depends on somewhat intractable cloud microphysical processes, discussed in chapter 4. If *all* the condensate drops out, then we can set $w_l = 0$ in (3.75). This yields the *pseudoadiabat*; 'pseudo' because it's not a real adiabat since entropy is not exactly conserved.

Some examples are shown in figure 3.7. To compute each of these curves, we start with a saturated parcel of specified temperature at the surface and integrate upwards. As the parcel rises, water condenses releasing latent heat and so temperature decreases more slowly than in the dry case. This effect is stronger the warmer (and hence moister) the initial conditions. At typical Earth-like surface temperatures, the effect is very strong: for a starting temperature of 5 °C, the mean lapse rate over the first 5 km is 7.2 K km^{-1}, for 20 °C it is 4.8 °C km^{-1} and for 35 °C it is 3.3 °C km^{-1} (compare with 9.8 °C km^{-1} for the dry adiabat).

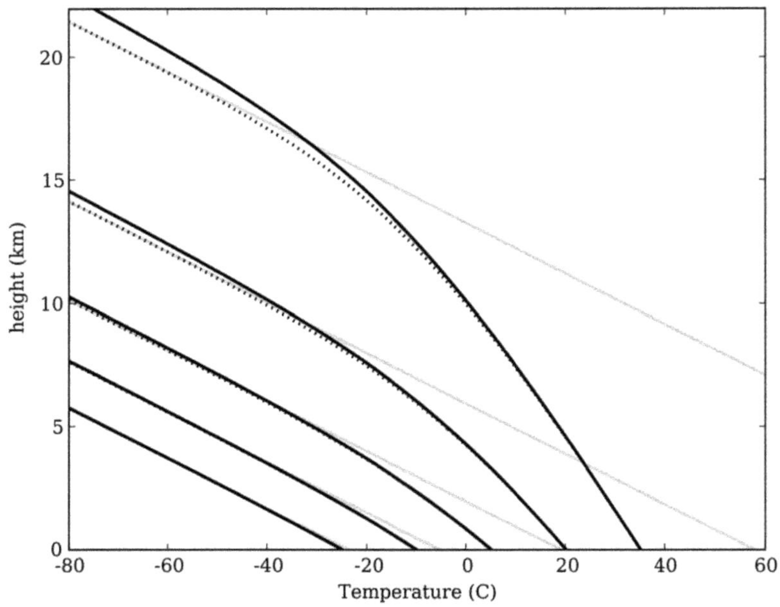

Figure 3.7. Moist adiabats (black solid lines) and pseudoadiabats (dotted lines) starting at $z = 0$ with temperatures of −25, −10, 5, 20 and 35 °C and the corresponding saturation humidity, with no condensed water initially. Surface pressure is 1000 hPa. The grey lines show dry adiabats to which the pseudoadiabats converge at high altitude.

As the parcel rises, more and more water condenses and the effect on the lapse rate becomes weaker. At great height, temperatures are very low, w_s becomes very small, and the pseudoadiabatic lapse rate converges to the dry adiabatic, g/c_{pd}, while the true adiabatic lapse rate converges to the smaller value $g/(c_{pd} + w_t c_l)$.

3.3.4 Visualizing the connection between the various meteorological temperatures

As we have seen, meteorologists enjoy defining a bewildering array of temperatures (potential, equivalent, wet-bulb, etc) which are connected to everyday absolute temperature by well-defined physical processes. These processes can be visualized graphically as shown in figure 3.8, which provides a handy way to tie all the temperatures together and remember the processes that connect them.

Consider a parcel initially at some height above the surface (black dot in figure 3.8), where it has some temperature T, some pressure p and some mixing ratio $w < w_s$ (i.e. it is unsaturated). If the parcel is lowered adiabatically, it will follow a dry adiabat. When it reaches the surface, its temperature will equal its potential temperature Θ (*potential temperature is the temperature an unsaturated parcel would have if brought adiabatically to the surface*). If the parcel is raised adiabatically, it will follow a dry adiabat up to the LCL. At this point the parcel is saturated and upon further lifting will follow a moist adiabat. If all condensate is removed from the parcel, then it follows a pseudoadiabat. Following a pseudoadiabat all the way to the top of the atmosphere results in complete drying of the parcel. If we then bring

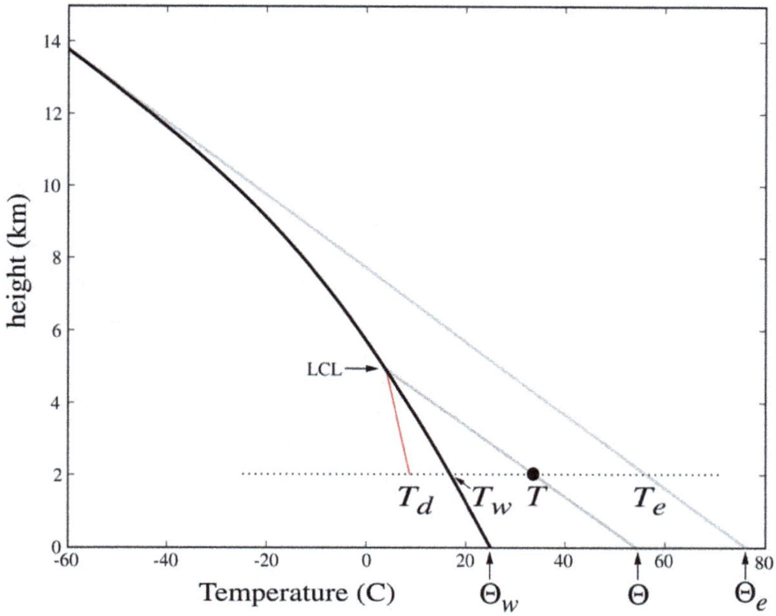

Figure 3.8. Transformations of an unsaturated parcel (black dot) lifted or lowered adiabatically from an initial state with temperature T. The black solid line is a moist pseudoadiabat. The grey lines are dry adiabats. The red line shows change in dew point temperature.

the parcel down again, it will follow a dry adiabat. When we get back down to the initial height, the parcel's temperature will equal the equivalent temperature T_e (*equivalent temperature is the temperature a parcel would have if all its water were made to condense while pressure was kept fixed*). If we keep going to the surface, the parcel's temperature will be the equivalent potential temperature Θ_e (*equivalent temperature is the temperature a parcel would have if all its water were made to condense and the parcel were brought adiabatically to the surface*).

Now consider the final possibility: we raise the parcel dry-adiabatically from its initial level to the LCL and then *lower* it pseudoadiabatically. In pseudoadiabatic descent, water is *added* to the parcel—just enough water to keep the parcel saturated (exactly the opposite to the removal of condensate in pseudoadiabatic ascent). When we get back to the initial level, the parcel's temperature will be the wet-bulb temperature T_w (*wet-bulb temperature is the temperature a parcel reaches by evaporating enough water to make itself saturated*). Finally, if we keep going to the surface (always following a pseudoadiabat) we reach the wet-bulb potential temperature Θ_w.

3.4 Static stability of a moist atmosphere with condensation

3.4.1 Conditional instability

In section 2.5.1, we examined the stability of a dry atmosphere to *infinitesimal* displacements of a test parcel. This gives two stability categories: if a parcel displaced upward becomes positively buoyant, then the temperature profile is unstable; otherwise, it is stable. When dealing with the stability of a moist atmosphere, it is useful to consider also *finite-size* displacements. This introduces a third category, called *conditional instability*: a profile is unstable if a parcel can become positively buoyant when displaced *far enough* upward.

Consider for instance the three idealized soundings depicted in figure 3.9. In all three cases, the ambient lapse rate is 7 °C km^{-1} and relative humidity is constant with height. In the left panel, relative humidity is 30%. A parcel lifted from the surface, being unsaturated, will initially follow a dry adiabat. Eventually, it will reach its LCL and then follow a moist adiabat. Throughout, its temperature will be less than that of the surroundings and the parcel will be negatively buoyant (i.e. will want to fall back down); the same is true for parcels lifted from any level. Thus the profile is stable to parcel displacements *of any size* from any level.

In the middle panel, relative humidity is 60%. A parcel lifted from the surface will initially be negatively buoyant. But some distance above the LCL, at a point called the *level of free convection* (LFC), the moist adiabat crosses the sounding temperature and the parcel becomes positively buoyant; this an example of conditional instability (which in turn is an example of *subcritical instability*, the general term for instabilities requiring a triggering perturbation of finite size). Parcels lifted from higher up in the atmosphere, on the other hand, never achieve positive buoyancy. Overall, the profile is stable to all infinitesimal perturbations but unstable to some finite-size perturbations, in this case, the profile as a whole is classed as conditionally unstable.

In the third case (right-hand panel) the atmosphere is saturated everywhere. Parcels near the surface are unstable even to infinitesimal perturbations, though parcels further up are stable; the profile as a whole is classed as unstable.

A word of warning: there is an alternative, and more traditional, definition of 'conditional instability', whereby a profile is conditionally unstable if its lapse rate is less than dry adiabatic but greater than moist adiabatic. This definition is fundamentally different from the finite-size perturbation definition given here: a profile that is conditionally unstable according to the traditional definition may actually be stable to all adiabatic parcel displacements of whatever size. Confusingly, the two

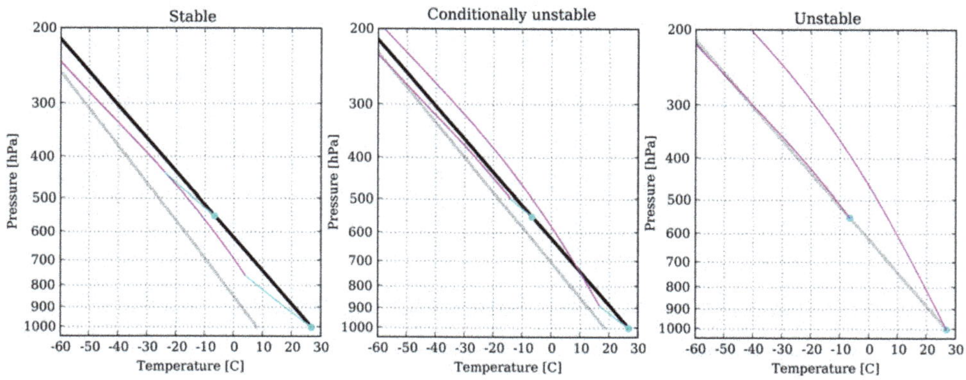

Figure 3.9. Three idealized temperature (black line) and dew-point (grey) soundings, all with the same fixed temperature lapse rate of 7 °C km^{-1} and a vertically constant relative humidity of 30% (left), 60% (middle) and 100% (right). Note that absolute and dew-point temperature coincide in the right-hand panel, so only the gray line is visible there.

definitions coexist and are sometimes mixed together. The conflict between the two definitions remains unresolved (see the discussion in Sherwood 2000). That such a basic definition should still be debated in a discipline over a century old is, among other things, a testament to the diversity of the those involved in meteorology, ranging from rough-and-ready practitioners to ivory-tower academic theorists. This diversity and the ensuing communication problems are both the bane and the charm of meteorology.

Now let's look at the stability of some real soundings. Figure 3.10 shows two examples, one from Kuching in Malaysia and the other from Valentia in Ireland. In the Kuching sounding, surface parcels reach their LFC at around 900 hPa and remain positively buoyant all the way up to about 120 hPa. Parcels higher up in the atmosphere are stable; overall, the sounding is conditionally unstable. The sounding at Valentia contains a strong *inversion* (a layer of the troposphere where temperature *increases* with height) which has a strongly stabilizing effect: parcels lifted from the surface experience some small positive buoyancy in a thin layer just under the inversion, but are negatively buoyant above the inversion. Overall, the profile may be classed as stable.

3.4.2 Skew-T and tephigram charts

Determining the stability of a sounding involves comparing the temperatures (and hence densities) of the sounding and of adiabatically lifted parcels. It is a fact of life that the troposphere is generally never *very* far from a moist adiabat—more precisely, the difference between a temperature sounding and a nearby moist adiabat is usually small compared with the overall temperature change from surface to tropopause. As a result, plots such as those along the top row in figure 3.10 are graphically inefficient: sounding and parcel trajectories are bunched up along the diagonal, with white space elsewhere. A neat trick to improve the presentation and make the important features stand out more clearly is to tilt the constant-temperature lines by 45°, as shown along the bottom row of figure 3.10: note how much more clearly (compared with the top row of the figure) you can see the regions of positive and negative buoyancy and how the shift from dry to moist adiabats is much more pronounced.

A *skew-T ln-p chart* is a special diagram used to plot atmospheric soundings. An example is shown in figure 3.11. Aside from pressure and tilted temperature lines, it has dry and moist adiabats at regular intervals, as well as lines showing dew-point temperature at fixed mixing ratio (so-called *mixing-ratio isopleths*; 'isopleth' means 'having the same value'). Since mixing ratio is conserved in an adiabatically lifted unsaturated parcel, these lines permit quick identification of the LCL: given the temperature and dew-point of a parcel, just follow temperature up the dry adiabat and dew-point up the humidity isopleth until the two meet and that's the LCL.

A *tephigram chart* is very similar to a skew-T chart, but it uses as coordinates Θ and T, with axes at right-angles[1]. Since $\Theta = T(p/p_0)^{-R_d/c_{pd}}$, a line of constant p is a

[1] Actually, the traditional choice of axis is T and $s = c_p \ln \Theta$, but since $\ln \Theta$ is essentially linear over the range of interest, it is simpler and almost equivalent to use Θ as an axis.

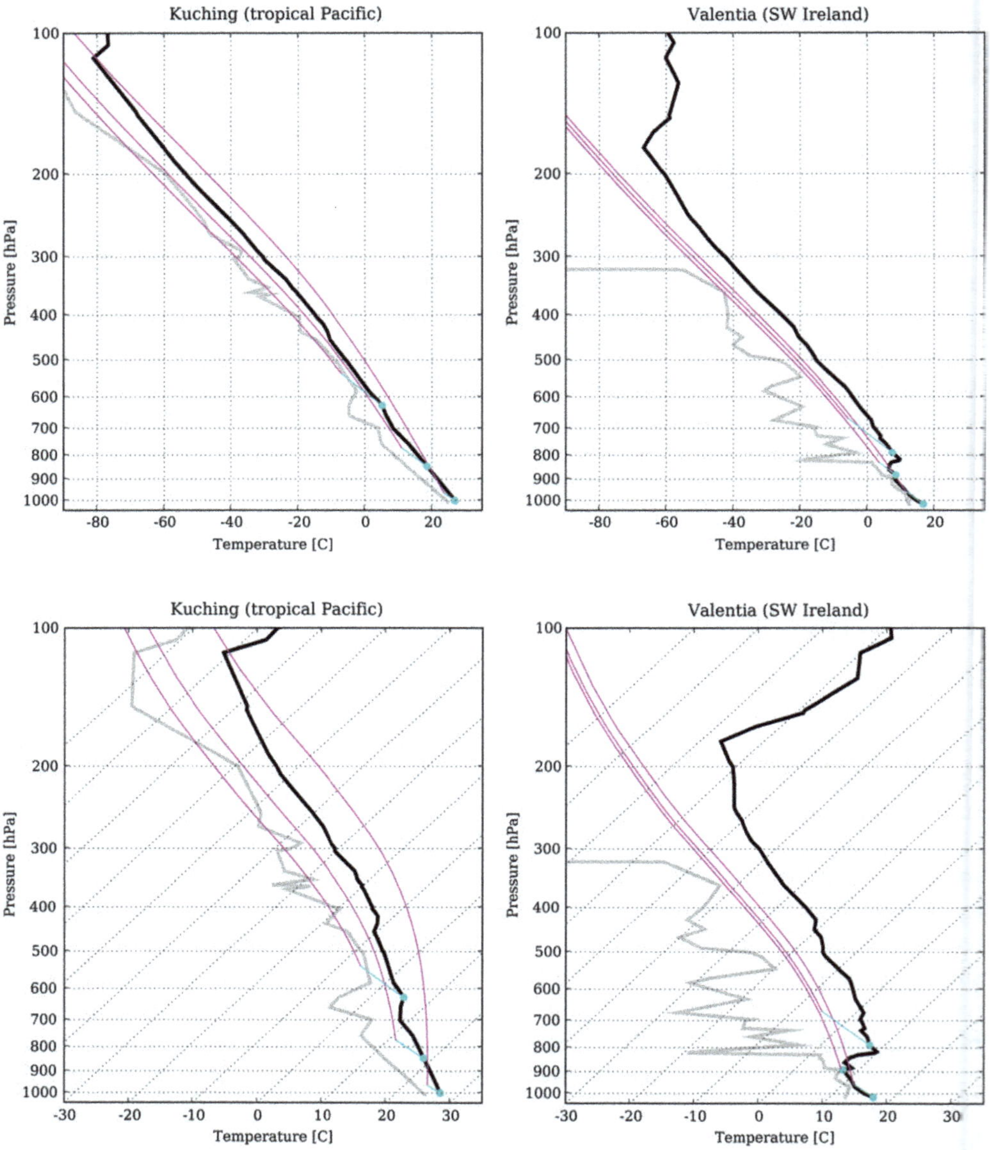

Figure 3.10. Real soundings from (left) Kuching, 6 Oct 2005 12Z, and (right) Valentia, 13 Oct 2006 12Z. The lower two panels show the same soundings plotted on a skew-T ln-p grid (the temperature isolines have been rotated clockwise by about 45°).

straight line making an angle $\arctan(p/p_0)^{-R_d/c_{pd}}$ to the horizontal. If we rotate this clockwise through 45°, then the $p = p_0 = 1000\,\text{hPa}$ line is horizontal, and lower-pressure lines are somewhat tilted. The result is very similar to a skew-T chart, but the dry adiabats are straight lines.

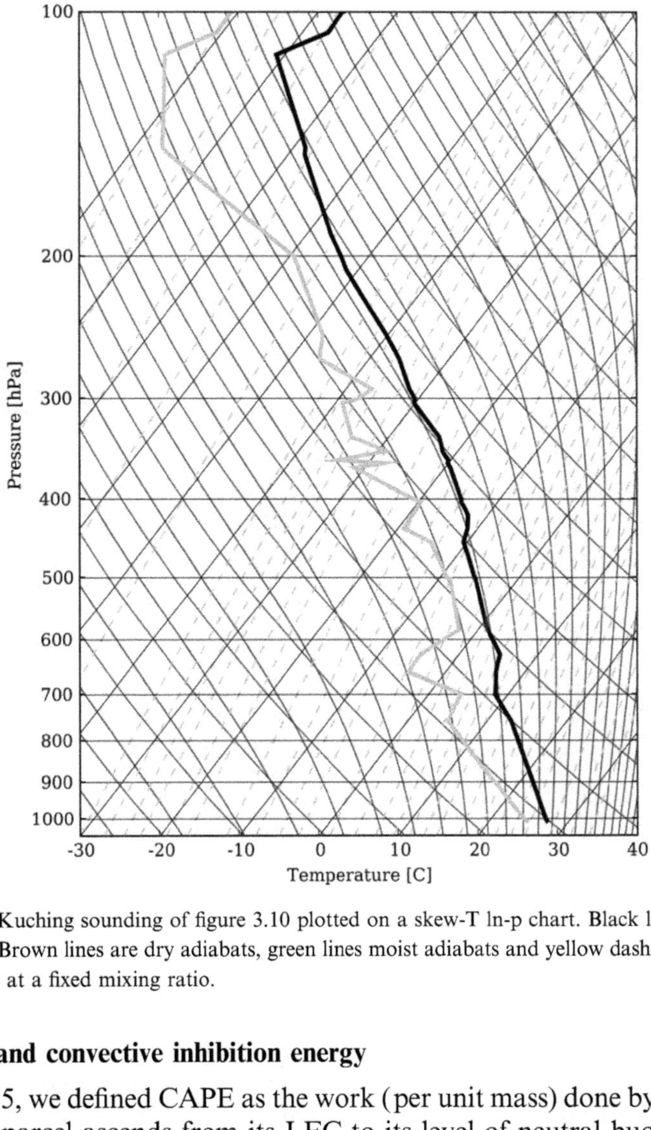

Figure 3.11. The Kuching sounding of figure 3.10 plotted on a skew-T ln-p chart. Black lines show pressure and temperature. Brown lines are dry adiabats, green lines moist adiabats and yellow dashed lines show dew-point temperature at a fixed mixing ratio.

3.4.3 CAPE and convective inhibition energy

In section 2.4.5, we defined CAPE as the work (per unit mass) done by the buoyancy force when a parcel ascends from its LFC to its level of neutral buoyancy (LNB). This carries over to the moist case:

$$\mathrm{CAPE} = R_\mathrm{d} \int_{p_\mathrm{LNB}}^{p_\mathrm{LFC}} \left(T_\mathrm{vp} - T_\mathrm{vs}\right) \mathrm{d} \ln p, \tag{3.78}$$

with the only difference that we use virtual temperature here. Graphically, this is the positive area between the parcel trajectory and sounding on a skew-T or tephigram plot. CAPE is positive for unstable and conditionally unstable parcels, and zero for stable parcels. The actual value of CAPE is related to the intensity of the ensuing convection.

Between the surface and the LFC, the parcel is negatively buoyant, so work must be done to lift it to the LFC. This is called the *convective inhibition energy* (CINE), given by

$$\mathrm{CINE} = R_\mathrm{d} \int_{p_\mathrm{LFC}}^{p_0} \left(T_\mathrm{vp} - T_\mathrm{vs}\right) \mathrm{d} \ln p, \tag{3.79}$$

which is always negative. The greater the CINE, the more work is needed to lift parcels to their LFC and the more difficult it is to trigger convection. The presence of CINE means that CAPE is not immediately released as soon as it is generated, but can accumulate until an adequate triggering event occurs. Very large amounts of CAPE can then be released all at once, leading to very intense storms.

3.4.4 Stability indices and thunderstorm forecasting

Because CAPE is somewhat complicated to compute, a number of simpler 'stability indices' have been devised over the years; these are numbers which, like CAPE, characterize the degree of instability of a profile and are used in forecasting severe weather (such as thunderstorms). The most classic is the *Showalter index* (SI), defined by

$$\mathrm{SI} = T_{500} - T'_{850} \tag{3.80}$$

where T'_{850} is the temperature of a parcel lifted pseudoadiabatically from 850 to 500 hPa, and T_{500} is the temperature of the surroundings at 500 hPa. Clearly, positive SI indicates stable conditions. Another widely used index is the *lifted index* (LI), defined by

$$\mathrm{LI} = T_{500} - T'_\mathrm{s} \tag{3.81}$$

where now T'_s is the temperature of a parcel lifted to 500 hPa from the surface. There are many other indices besides these two, all of which compare the temperature, humidity and winds at different levels.

Stability indices are indicators of convection and thunderstorms, but only in a statistical sense. Thus, given a profile with very negative SI (or LI), there is no guarantee that convection will occur, but there is a *greater probability* that it will. For stability indices to be quantitatively useful in forecasting, we need to know the *conditional probability* that convection will occur given that the index has a certain value (or, equivalently, the probability that the index has a certain value given that convection is observed). These probabilities can be estimated empirically, by computing the index for a great many soundings at a given site and observing whether or not a thunderstorm develops. Once we have these probability distributions, we can make a probabilistic forecast: given a value of SI, we can say what is the probability that a thunderstorm will develop today. The catch is that the probability distributions are site-specific, so we cannot use data from one station to make predictions about faraway locations. In general, stability indices are used in a 'fuzzy' way, as a qualitative indicator. Operational forecasters, after a few years of hands-on work, develop a seat-of-the-pants feel for how likely a thunderstorm is given SI and other data.

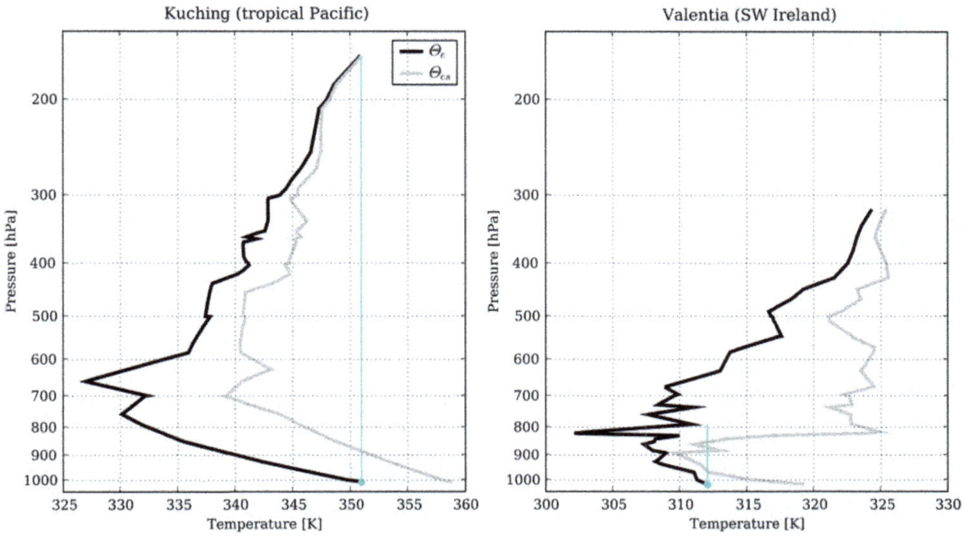

Figure 3.12. Θ_e and Θ_{es} for the soundings shown in figure 3.10. The blue line shows the (constant) Θ_e of a parcel lifted from the ground.

3.4.5 Relation between Θ_e, Θ_{es} and stability

We saw in section 2.5.2 that the static stability of dry air is determined by the vertical rate of change of Θ: if Θ increases with height, the profile is stable, otherwise it is unstable. This is because *at a given pressure*, Θ depends only on T; so if we bring two parcels adiabatically to the same pressure, the one with higher Θ will always be warmer and thus lighter than the other.

For a parcel that can undergo condensation, things get more complicated. The conserved quantity in this case is Θ_e, which depends on humidity as much as on temperature. Thus, two parcels at the same pressure and with the same Θ_e can have very different temperatures; there is no simple relation between Θ_e and parcel buoyancies. However, we can get around this by comparing the Θ_e of the test parcel with the *saturated* Θ_e (i.e. Θ_{es}) of the environment. By making the environment saturated, we remove the dependency on humidity, so that the comparison reflects only differences in temperatures. It can be shown that *for a parcel that has achieved saturation* (i.e. one that is above its LCL) the approximate relation

$$\Theta' - \Theta \simeq \frac{\Theta_e' - \Theta_{es}}{1 + \beta} \tag{3.82}$$

is valid, where primes refer to the adiabatically lifted parcel and $\beta = (\ell_v/c_p)\partial w_s/\partial T$. Figure 3.12 shows this for the two real soundings.

Exercise 3.4.5: Show that (3.82) is true. To do so, expand the exponential in Θ_e and Θ_{es} to first order in a Taylor series, take the difference and use a further expansion on $w_s(T)$.

3.5 Mixing clouds and contrails

To form condensation in air, i.e. to make a cloud, you need to either cool the air, make it more moist, or both. The easiest, fastest and by far the most common way for air to cool sufficiently to form a cloud is by adiabatic expansion; hence the importance of lifting and the emphasis on adiabatic processes in all we've done up to here. But there are other ways to cool air and, though they play a less important role, they are worth discussing. These processes involve exchange of heat and/or mass between an air parcel and its surroundings, so they are *diabatic* (some prefer the term 'non-adiabatic'; and why not a-adiabatic or a-nondiabatic?).

One example is *radiative cooling*. As we will see later, all bodies (including bodies of gas) spontaneously emit radiation, which carries away energy. Under the right conditions (at night and under clear skies with weak winds), the Earth's surface can cool dramatically through radiative loss (hence the cold desert nights we're all familiar with from films such as Lawrence of Arabia). The air in contact with the surface will cool by conduction and, if it is humid enough, a fog will form; this is called *radiation fog*.

Another way to cool air is by mixing it with colder air. This is what happens when you can see your breath on a cold day: the warm moist air coming out of your mouth mixes with colder ambient air and, if conditions are right, the resultant mix is supersaturated so condensation forms. A more dramatic example of the same phenomenon are *contrails* (short for *condensation trails*), the linear clouds stretching behind aeroplanes high in the sky (figure 3.13, left). Contrails are of interest to various people, among them the Air Force: it just won't do for a multi-zillion-dollar stealth fighter to have bright white cloud trailing behind it. Interest in contrails also derives from their possible role in enhancing global warming. If you've ever stopped to look at contrails for a while, you will have noticed that often they don't just disappear, but evolve into more horizontally extensive cirrus clouds. These clouds

Figure 3.13. (Left) A jet liner producing contrails. (Right) An image from the NASA MODIS satellite showing extensive contrail coverage over the southeastern United States. Image credit: NASA.

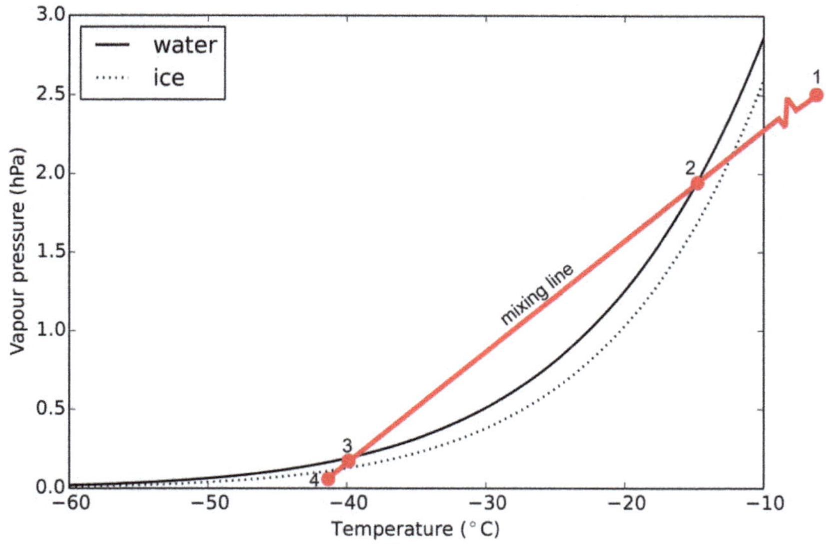

Figure 3.14. A schematic of contrail formation.

are thin enough to let plenty of sunlight through, but they still trap infrared radiation quite strongly (see chapter 5), so they lead to a net warming. In regions of the world with heavy air traffic, contrail and associated cirrus coverage can become very extensive (figure 3.13, right.)

To understand contrail formation, it helps to look at a figure like 3.14. A jet engine is a machine that takes air with ambient temperature and humidity (represented by point 4 in the figure), adds heat and water vapour to it (both resulting from combustion) and spews it back out again in a highly turbulent state (point 1). Because of the turbulence, the hot, moist exhaust air rapidly mixes with the much colder ambient air. If mass m_e of exhaust air mixes with mass m_a of ambient air, the temperature of the mixture is $T = (1 - f)T_a + f T_a$, where T_e is exhaust temperature, T_a is ambient temperature, and $f = m_a/(m_e + m_a)$ is the mixing fraction. A similar expression is valid for the humidity. As the exhaust air becomes more and more diluted, f increases and the point representing the state of the mixed air moves from point 1 to point 4 along a straight line called the *mixing line*. If the mixing line crosses the saturation vapour pressure curves, then a contrail will form. Normally, the ambient air is unsaturated, and so the contrail will eventually dissipate as the mixed air approaches point 4 (this is why contrails typically have a beginning and an end). However, it can happen that point 4 lies in between the ice and the water saturation curves (i.e. the air is supersaturated with respect to ice but unsaturated with respect to water). Under these conditions, clouds will not form spontaneously (because it is difficult to nucleate ice drops directly), but icy contrails will persist for a long time.

Exercise 3.5: The air coming out of a jet engine is much warmer than its surroundings and therefore very buoyant; you would expect it to shoot up into the sky like a balloon. However, this does not happen appreciably: contrails form roughly at the same level as the aeroplane. Give a quantitative (order-of-magnitude) explanation for this. It helps to look closely at figure 3.13 and make reasonable assumptions about the speed and length of the aeroplane and the temperature difference between the exhaust and ambient air.

Bibliography

Sherwood S C 2000 On moist instability *Mon. Weather Rev.* **128** 4139–42

IOP Publishing

Physics of the Atmosphere

Rodrigo Caballero

Chapter 4

Cloud microphysics

4.1 Homogeneous nucleation

How is a cloud formed? Well, you take a parcel of moist air, lift it adiabatically beyond its LCL, the air becomes saturated and the excess water condenses out; end of story. Right? Life is never so simple, unfortunately (or fortunately, depending on your viewpoint: if life were simple, we wouldn't need scientists, or universities). This simple picture of cloud formation, empirically supported by the observation that supersaturation greater than a few percent is never observed in the atmosphere, actually conceals a considerable amount of complexity related to the details of the way condensation forms.

Take an enclosed parcel of *pure* moist air (i.e. just dry air and water vapour, no other junk) and cool it to its dew point. What happens? As it turns out, nothing. No cloud forms. In fact you need to cool the air to the point where it is supersaturated by several hundred percent before a cloud will form. To understand this, we need to consider the process by which a cloud drop is formed. To form a cloud droplet out of pure moist air, you need to start from a tiny clump of water molecules which come together by chance. This proto-droplet, though tiny, is nevertheless liquid. As with any other liquid surface, we can define a saturation vapour pressure as the vapour pressure such that the flux of molecules entering the liquid matches the outgoing flux (section 3.2.2). If the ambient vapour pressure is greater than the saturation value, then the flux of molecules entering the drop exceeds the number leaving. The droplet grows, and continues growing so long as the ambient air remains supersaturated. This process is called *homogeneous nucleation*.

The catch is that *saturation vapour pressure depends on droplet radius*: the smaller the drop, the greater the vapour pressure required to keep it in equilibrium. Thus, in the thought experiment above, the air is supersaturated with respect to a *flat* surface of water (infinite radius), but is subsaturated with respect to the tiny proto-drops generated by chance, which evaporate as soon as they are formed. It is only when supersaturation (with respect to a flat surface) exceeds 400% that a significant number of proto-drops can exist in equilibrium.

The reason for the radius dependence is that water molecules find it a lot easier to escape from the surface of small drops than they do from a flat surface. This can be seen as a geometric effect. Inter-molecular forces act over a characteristic distance; when droplet radius becomes comparable to this distance, a water molecule just above the surface of the droplet will in effect interact with (and be attracted to the liquid phase by) a much smaller number of molecules than if the liquid surface were flat. As a result, it takes less work for a molecule to leave a small droplet than a large one and the resulting saturation vapour pressure is consequently higher for the small droplet.

4.1.1 Kelvin's equation

These ideas can be quantified by introducing the concept of *surface tension*, which roughly speaking measures the force required to stretch the surface 'skin' of a liquid (the existence of this skin is exploited by insects light enough to walk on water without punching through its surface). As a droplet evaporates, its skin shrinks, resulting in a release of PE which gives an increasingly important contribution to the overall energetics of the evaporation/condensation process the smaller the radius. Though the derivation is not particularly more difficult than anything we've done before, we will omit it for brevity. The result is *Kelvin's equation*, which gives the radius dependence of saturation vapour pressure:

$$e_s(T, r) = e_{s\infty} \exp\left(\frac{2\sigma}{rR_v\rho_l T}\right), \quad (4.1)$$

where $e_s(T, r)$ is the saturation vapour pressure above a drop of radius r and temperature T, $e_{s\infty}$ is the saturation vapour pressure above a flat surface, σ is the surface tension and ρ_l is the density of liquid water.

Kelvin's equation has a simple physical interpretation. Recalling expression (3.39) for $e_{s\infty}$, we can write

$$e_s(T, r) = C \exp\left(-\frac{m_v \ell_v - 2\sigma m_v / \rho_l r}{kT}\right), \quad (4.2)$$

where C is a constant and m_v is the mass of a water molecule. The quantity in the exponent is the ratio of the energy required for a molecule to break free from the droplet to the mean KE of the molecules. The energy required to escape it is $m_v \ell_v$ if the surface is flat ($r = \infty$), but decreases as the radius decreases.

A typical value for σ is 7.3×10^{-2} N m^{-1}. Using this value, we can plot e_s as a function of r (figure 4.1). Recall that is e_s is the vapour pressure at which a droplet of given radius and temperature can exist in equilibrium with its environment—that is, $e_s(r, T)$ is the vapour pressure such that a droplet of radius r and temperature T will *in principle* neither grow nor shrink, but maintain its size indefinitely. However, this is an *unstable* equilibrium. Droplet radius is subject to continuous fluctuations,

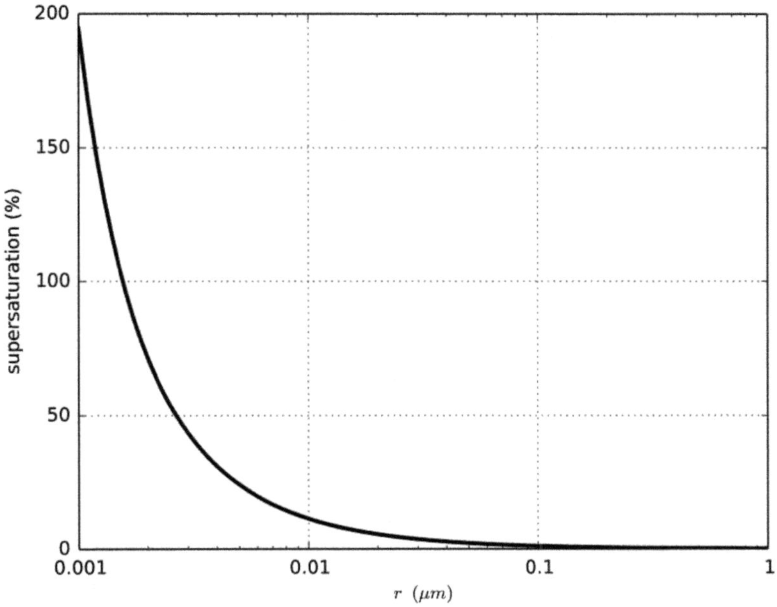

Figure 4.1. Kelvin's equation: supersaturation $e_s(r, T)/e_{s\infty}(T) - 1$ computed using (4.1) at $T = 20$ °C.

due to the randomness of the processes by which molecules evaporate and condense onto it. If, by a random fluctuation, droplet radius increases, this will make it more difficult for molecules to escape from it, reducing the average outgoing molecule flux. At the same time, the average incoming flux remains unchanged, so the drop will gain even more mass and so on in a runaway process. Similarly, a small negative fluctuation away from equilibrium will cause the drop to evaporate and disappear altogether. Thus *in practice* a small droplet will not be observed to maintain its radius indefinitely: it will either grow or disappear.

4.2 Heterogeneous nucleation and aerosols

Since supersaturation never exceeds a few percent in the atmosphere, homogeneous nucleation cannot play any role. So how do clouds actually form? The answer is *heterogeneous nucleation*: all cloud droplets start their life by condensing around a speck of foreign matter. All real air contains a large number of tiny particles. When such particles are small enough, Brownian motion overwhelms gravity and the particles can remain in suspension indefinitely, in which case they are called *aerosols*. Though small, each aerosol particle still contains enough molecules to be considered a chunk of solid or liquid (i.e. an aggregated state of matter). When they get small enough, aerosols start to behave more like a true solution; this happens when they are smaller than about 10^{-3} μm. The opposite extreme, in which gravity overwhelms Brownian motion, is called a *suspension*; suspensions will eventually settle out given sufficient time. The boundary between aerosols and suspensions is fuzzy, but is generally set at around 1 μm. Clouds are an example of a suspension. Both aerosols and suspensions form part of the broader class of systems known as *colloids*.

The origin and evolution of aerosols is a very complicated topic in atmospheric chemistry. The chemical composition of aerosols is very varied—anything from very fine mineral dust, to soot, to sulphate and nitrate salts. Human activities, especially fossil fuel burning, produce a lot of aerosols. In general, aerosols are much more abundant over continents than over the oceans, which has consequences for the clouds that form there. Because aerosols interact with radiation and are crucial to cloud formation, they can strongly affect climate.

Not all aerosol particles can be used to form a cloud droplet; certain criteria apply. Those aerosols meeting these criteria are called *cloud condensation nuclei* (CCN). There are two types of aerosol which can act as CCN.

1. The aerosol is insoluble but *wettable* (or *hydrophilic*) and sufficiently large. A surface is wettable when water collects on it in a thin film rather than bunching into drops (the opposite of a wettable surface is called *hydrophobic*, the classic example being a well-waxed car). In a moist atmosphere, a wettable aerosol will collect a film of water around itself and act, to all intents and purposes, as a droplet of pure water. If the particle's radius is large enough, this 'drop' will find itself to the right of the Kelvin equilibrium line and will grow indefinitely: a cloud drop is born. However, for this to happen at supersaturations of only a few percent requires aerosols of radius ~1 μm. Aerosols this big are rare, so this mechanisms is not the most common way to form cloud drops.
2. The other way for an aerosol to act as a CCN is for it to be *soluble* and large enough. Though there is still a size restriction, it is much less stringent than in the previous case. Because there are many more small aerosols than larger ones, this is the most common way for cloud droplets to form. We examine this mechanism in detail in the following section.

4.2.1 Raoult's law

Vapour pressure is also affected by the purity of the water. This dependence is described by *Raoult's law*:

$$e_s(T, r, f) = f\, e_s(T, r, 1) \tag{4.3}$$

where f is the number fraction of water molecules in the solution:

$$f = \frac{n_w}{n_w + i\, n_s}, \tag{4.4}$$

with n_w the number density of water molecules, n_s the number density of solute molecules, and i the number of ions into which the solute dissolves (e.g., $i = 1$ for ethanol, which does not ionize, while $i = 2$ for sodium chloride, which separates into Na^+ and Cl^- ions). $e_s(T, r, f)$ is the water vapour pressure above a drop at temperature T, radius r and concentration f; $e_s(T, r, 1)$ is the vapour pressure above a similar drop of pure water. Note that since $f \leqslant 1$, adding impurities always *reduces* vapour pressure.

Exercise 4.2.1: A sling psychrometer (section 3.2.5) is used to measure humidity. Estimate the error in the humidity measurement if the water used in the wet bulb is dirty, so that of every ten molecules, only nine are water molecules.

Raoult's law is striking both for its simplicity and for the fact that it applies to *any* solvent–solute combination: it depends only on concentration and not on the specific chemical species involved. It is thus extremely powerful. The downside is that the law is not actually true in general. It is exactly true only for *ideal solutions*, i.e. those for which there is no change in temperature or volume upon mixing solvent and solute. However, it also applies approximately to non-ideal solutions provided they are *dilute* ($1 - f \ll 1$). Physically, the reduction in vapour pressure has nothing to do with attraction between solvent and solute molecules, as you might think—in fact, an ideal solution can be defined as one in which such inter-molecular forces are zero. Rather, the effect of the solute molecules is to take up space at the surface of the liquid, making it harder for water molecules to find their way out of the liquid. The fraction of surface area taken up by solute molecules is proportional to their concentration, which explains the simple form of (4.4).

4.2.2 Köhler diagrams

Now consider a soluble aerosol suspended in a moist atmosphere. Water molecules will condense onto the aerosol, since initially $f \ll 1$ and so the atmosphere is strongly supersaturated with respect to the aerosol. Eventually the aerosol becomes a droplet of solution. The saturation vapour pressure above such a drop is obtained by combining Raoult's law (4.3) with Kelvin's equation (4.1), which gives the *Köhler relation*

$$e_s(T, r, M_s) = \frac{e_{s\infty}}{1 + AM_s/r^3} \exp\left(\frac{B}{rT}\right) \tag{4.5}$$

$$\simeq e_{s\infty}\left(1 - \frac{AM_s}{r^3} + \frac{B}{rT}\right), \tag{4.6}$$

where $e_{s\infty}$ is the saturation vapour pressure above a flat surface of *pure* water,

$$A = \frac{3im_v}{4\pi\rho_l m_s} \qquad B = \frac{2\sigma}{R_v \rho_l}, \tag{4.7}$$

M_s is the initial mass of the aerosol, m_s the aerosol's molecular mass, m_v is the molecular mass of water and ρ_l is the density of liquid water.

Exercise 4.2.2: Derive (4.5).

Example Köhler curves for a range of aerosol sizes are shown in figure 4.2. At small r, the cubic solute effect dominates and the equilibrium vapour pressure is much less than for pure water. Thus, small solution droplets can comfortably exist in

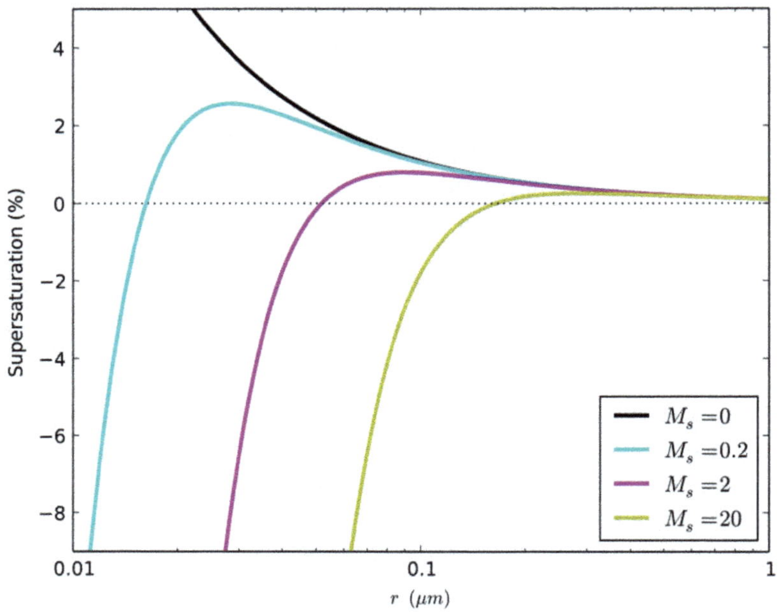

Figure 4.2. A Köhler diagram computed using (4.5) for sodium chloride aerosols with the masses M_s indicated (in units of 10^{-20} kg). The temperature is 20 °C. The molecular mass of NaCl is 58.44 AMU. The ordinate shows supersaturation with respect to a flat surface of pure water, $e_s(T, r, M_s)/e_{s\infty}(T) - 1$.

equilibrium even at relative humidities less than 100% (i.e. $e_s/e_{s\infty} < 1$). Below 100% relative humidity, a single equilibrium radius is possible (for a given aerosol size) and this is a *stable* equilibrium: if a small fluctuation makes r suddenly larger, the drop finds itself subsaturated and evaporates until it regains its equilibrium radius. These stable droplets are called *haze droplets* and are responsible for the hazy, milky appearance of the horizon on a warm, humid day.

At relative humidities greater than 100%, two equilibrium radii are possible. The left-hand equilibrium is stable, while the right-hand equilibrium is unstable, and drops will grow indefinitely. For each aerosol size, there is a *critical radius* which separates stable from unstable equilibria. The critical radius corresponds to the highest relative humidity at which droplets can exist in equilibrium: at higher relative humidities, droplets will grow indefinitely. This, then, is how most cloud droplets are born. As air is cooled and relative humidity increases, haze droplets increase steadily in size following the Köhler curve. Once the critical radius is reached, the droplets continue to grow spontaneously. Having reached this stage, droplets are said to be 'activated' and they will continue to grow so long as the supersaturation is greater than zero (note that activated droplets continue to grow even if the supersaturation drops below the activation level).

4.3 Droplet growth by condensation

Air parcels in nature normally contain a large number of aerosols with a range of sizes. When a parcel is adiabatically lifted, its relative humidity increases (section 3.2.8) and some of the aerosols become haze droplets. Soon after the parcel is lifted beyond its LCL, the largest of the haze droplets will be activated. As the parcel is lifted further and supersaturation increases, smaller and smaller haze droplets will be activated. The activated droplets grow by condensation, removing moisture from the air and counteracting the tendency for supersaturation to increase as the parcel is lifted. At some point, so many droplets are activated that the two effects balance, supersaturation no longer increases and no new haze droplets are activated. Understanding the processes controlling the evolution of droplet size and supersaturation in such an adiabatically lifted parcel is a classical problem in atmospheric physics. It is usually tackled by first examining the growth of an individual droplet and then considering a population of droplets and their feedback on the ambient air.

4.3.1 Growth of an isolated droplet

Consider a droplet instantaneously inserted into supersaturated air; throughout this discussion we will ignore curvature and solute effects, so we define supersaturation as

$$s = \frac{e}{e_{s\infty}(T)} - 1 \tag{4.8}$$

where e is the actual vapour pressure of the ambient air and T is its temperature. Suppose that the ambient moisture and temperature fields are initially uniform. The droplet is out of equilibrium with its supersaturated environment and water molecules will enter the droplet at a higher rate than they exit. Humidity near the droplet will be depleted, creating a moisture gradient with the unchanged humidity farther afield. Molecular diffusion will act on this gradient, delivering further moisture to the droplet so that condensation may continue. At the same time, the droplet will warm up due to latent heat release; this will build up a reverse temperature gradient so that thermal diffusion will carry heat away from the droplet. Eventually, a steady state will be reached in which the droplet has a constant temperature somewhat higher than the environment (but still low enough to be supersaturated) and there is a constant flux of moisture into the droplet and a constant outward flux of energy.

To formalize this picture we assume that the transport of humidity and energy are both governed by *diffusion*. A physical quantity Q it is said to behave diffusively if its flux is proportional to its concentration gradient:

$$\mathbf{F} = -D\nabla c \tag{4.9}$$

where $\mathbf{F}$ is the flux (amount of Q crossing unit area in unit time), c is the concentration (amount of Q per unit volume) and D is the *diffusivity*, a constant of proportionality with units of [length]^2/[time]. Note that the flux is *down* the concentration gradient. The local rate of change of concentration will then be simply the convergence (negative divergence) of the flux, or

$$\frac{\partial c}{\partial t} = D\nabla^2 c, \tag{4.10}$$

where we have assumed that D is spatially uniform. We further assume that the problem is spherically symmetric, so can focus on the radial component

$$\frac{\partial c}{\partial t} = \frac{D}{R^2}\frac{\partial}{\partial R}\left(R^2\frac{\partial c}{\partial R}\right) \quad (4.11)$$

where R is the radial distance from the droplet centre.

Now suppose that the droplet has radius r metres, and we take this as the unit of length. We also take an arbitrary period τ seconds as the unit of time. We can then rewrite (4.11) as

$$\frac{r^2}{\tau D}\frac{\partial c}{\partial t} = \frac{1}{R^2}\frac{\partial}{\partial R}\left(R^2\frac{\partial c}{\partial R}\right) \quad (4.12)$$

where now R is in units of r and t is in units of τ. Plausibly, the left-hand side can be neglected when $r^2/\tau D \ll 1$, which requires $\tau \gg r^2/D$. Under typical atmospheric conditions, the diffusivity of both moisture and temperature is on the order of $10^{-5}\,\mathrm{m^2\,s^{-1}}$ while typical cloud drops have a radius of 10^{-5} m. Over distances of a few droplet radii, the system can therefore be considered to be in a steady state after $\sim 10^{-4}$ s. In other words, the humidity and temperature structures in the vicinity of the droplet equilibrate essentially instantaneously with the much slower changes driven by adiabatic ascent and by the growth of the droplet itself. We need therefore only focus on the steady-state equation

$$\frac{1}{R^2}\frac{\partial}{\partial R}\left(R^2\frac{\partial c}{\partial R}\right) = 0. \quad (4.13)$$

Taking boundary conditions $c = c_\mathrm{r}$ at the droplet surface and $c = c_\infty$ very far from the droplet, the solution is

$$c(R) = c_\infty - (c_\infty - c_\mathrm{r})\frac{r}{R}. \quad (4.14)$$

As shown in figure 4.3 this profile has a boundary-layer character, with sharp gradients concentrated near the droplet while the concentration field remains largely undisturbed at distances greater than about ten radii. This is a geometric effect: from (4.9), the total flux across a spherical surface concentric with the droplet is

$$-4\pi r D(c_\infty - c_\mathrm{r}) \quad (4.15)$$

which is independent of R; a small sphere thus requires a greater gradient to support the same total flux as a larger sphere.

We are specifically interested in the flux of water vapour into the drop and of energy away from it. For the former, the relevant concentration is the mass density of water vapour ρ_v and using the ideal gas law we set $c_\infty = e/R_\mathrm{v}T$, where e and T are the ambient vapour pressure and temperature while R_v is the gas constant for water vapour, and $c_\mathrm{r} = e_\mathrm{s}(T_\mathrm{r})/R_\mathrm{v}T_\mathrm{r}$ where e_s is the saturation vapour pressure and T_r is the

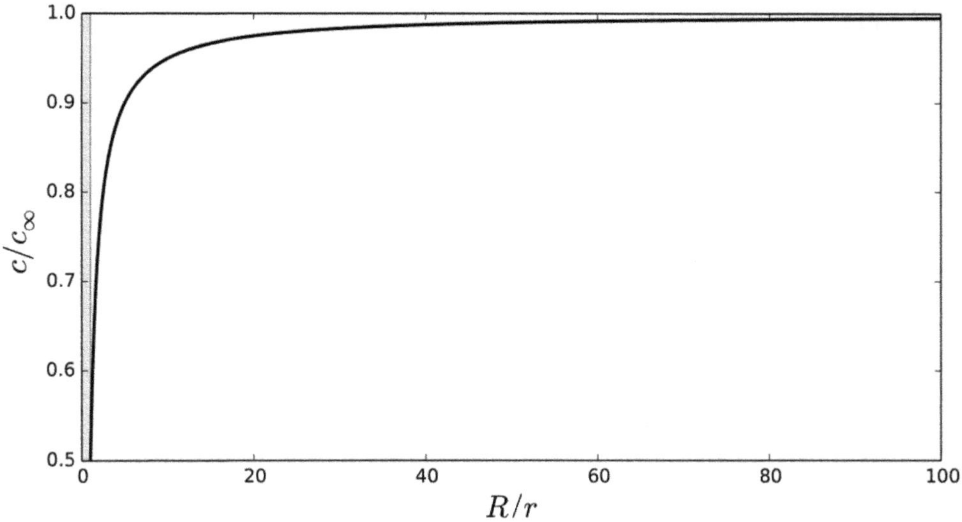

Figure 4.3. A plot of the steady-state diffusive profile (4.9) for a case in which $c_r/c_\infty = 0.5$. Grey shading denotes the droplet itself.

droplet temperature which we assume to be uniform[1]. The rate of change of droplet mass dm/dt must be equal to the total inward flux of water vapour across the surface:

$$\frac{dm}{dt} = 4\pi r D\left(\frac{e}{R_v T} - \frac{e_s(T)}{R_v T_r}\right). \tag{4.16}$$

For energy transport, the relevant concentration field is the enthalpy per unit volume $h = \rho c_p T$, where ρ is the air density (see section 2.3.8). In the steady state, the outward energy transport must match the latent heating, so that

$$\ell_v \frac{dm}{dt} = 4\pi r K (T_r - T) \tag{4.17}$$

where $K = \rho c_p D$ is called the *thermal conductivity* of air.

Equations (4.16) and (4.17) together with the Clausius–Clapeyron relation (3.37) form a complete set which can be solved to yield T_r and dm/dt given the ambient temperature and humidity but, since the Clausius–Clapeyron relation involves transcendental functions, this solution must be obtained numerically. We can, however, obtain an approximate analytical solution by noting that supersaturation values in the atmosphere are always very small, which implies $(T_r - T)/T \ll 1$. We can then rewrite (4.16) approximately as

$$\left(\frac{R_v T}{4\pi r D}\right)\frac{dm}{dt} = e - e_s(T_r). \tag{4.18}$$

[1] In reality, latent heat is deposited at the droplet surface and must be carried inwards by diffusion, but the time scale for this process is again fast enough to be considered instantaneous.

Furthermore, using a first-order Taylor expansion and the Clausius–Clapeyron relation we have

$$e_s(T_r) - e_s(T) \approx \frac{e_s(T)\ell_v}{R_v T^2}(T_r - T) \tag{4.19}$$

so we can rewrite (4.17) as

$$\left(\frac{e_s(T)\ell_v^2}{4\pi r K R_v T^2}\right)\frac{dm}{dt} = e_s(T_r) - e_s(T). \tag{4.20}$$

Finally, we add (4.18) to (4.20) and write the droplet mass as $m = 4\pi\rho_l r^3/3$, where ρ_l is the density of liquid water, to obtain

$$\frac{dr}{dt} = A\frac{s}{r} \tag{4.21}$$

where s is the supersaturation (4.8) and

$$A = \left(\frac{\rho_l R_v T}{D e_s(T)} + \frac{\rho_l \ell_v^2}{K R_v T^2}\right)^{-1}. \tag{4.22}$$

A depends weakly on temperature and pressure and has a typical value of around 50 $\mu m^2\ s^{-1}$. The key result here is that the radius growth rate dr/dt is inversely proportional to r itself: *small droplets grow faster than large ones.*

4.3.2 Growth of a population of droplets

We turn now to the problem of how a collection of droplets evolves in an adiabatically lifted parcel. An approximation that is invariably made in tackling this problem is to assume that the droplets all interact with an ambient supersaturation field which is spatially uniform and can be described by a single number. This assumption is partly justified by the observation that typical droplet separations are on the order of 1 mm or around 100 typical droplet radii; droplets will therefore not generally 'feel' the deformation of the supersaturation field induced by other droplets (see figure 4.3). Nonetheless, clouds contain a mixture of different droplet sizes distributed randomly in space and we expect the supersaturation to decrease more quickly in areas that happen to be rich in larger droplets—note in fact that (4.21) implies faster mass growth for larger particles (even though their radius grows more slowly) and hence faster depletion of ambient supersaturation. We must therefore assume that such effects are negligible or that large-scale diffusion or another mixing mechanism keeps the supersaturation reasonably uniform.

With this assumption, the question is to understand how adiabatic lifting and condensation affect the supersaturation. We can write

$$\frac{ds}{dt} = \left.\frac{ds}{dt}\right|_{\text{lift}} + \left.\frac{ds}{dt}\right|_{\text{cond}} \tag{4.23}$$

where the first term on the right-hand side assumes no condensation and the second term assumes no lifting.

For the lifting term, we have

$$\left.\frac{\mathrm{d}s}{\mathrm{d}t}\right|_{\text{lift}} = \frac{\mathrm{d}s}{\mathrm{d}z} v_z \tag{4.24}$$

where v_z is the vertical speed of the parcel and

$$\frac{\mathrm{d}s}{\mathrm{d}z} = \frac{1}{e_s}\frac{\mathrm{d}e}{\mathrm{d}z} - \frac{e}{e_s^2}\frac{\mathrm{d}e_s}{\mathrm{d}z}. \tag{4.25}$$

If there is no condensation then the partial pressure of water vapour decreases at the same rate as the total pressure, which assuming hydrostatic balance means

$$\frac{\mathrm{d}e}{\mathrm{d}z} = -\frac{eg}{RT}. \tag{4.26}$$

Temperature decreases at roughly the dry adiabatic rate (see section 2.4.7), so

$$\frac{\mathrm{d}e_s}{\mathrm{d}z} = \frac{e_s \ell_v}{R_v T^2}\frac{\mathrm{d}T}{\mathrm{d}z} = -\frac{e_s \ell_v g}{R_v T^2 c_p}. \tag{4.27}$$

Turning to the condensation term, we have

$$\frac{\mathrm{d}e}{\mathrm{d}t} = R_v T\frac{\mathrm{d}\rho_v}{\mathrm{d}t} = -R_v T\frac{\mathrm{d}M}{\mathrm{d}t} \tag{4.28}$$

where M is the total mass density of condensate, while

$$\frac{\mathrm{d}e_s}{\mathrm{d}t} = \frac{\mathrm{d}e_s}{\mathrm{d}T}\frac{\mathrm{d}T}{\mathrm{d}t} = \frac{e_s \ell_v}{R_v T^2}\frac{\ell_v}{\rho c_p}\frac{\mathrm{d}M}{\mathrm{d}t}. \tag{4.29}$$

Using (4.21), the total condensation rate can be written as

$$\frac{\mathrm{d}M}{\mathrm{d}t} = 4\pi\rho_1 N\bar{r}As \tag{4.30}$$

where N is the number of droplets per unit volume and $\bar{r}$ is the mean droplet radius.

Putting it all together and taking $e/e_s \approx 1$ we finally obtain

$$\frac{\mathrm{d}s}{\mathrm{d}t} = B\,v_z - C\,\bar{r}\,s \tag{4.31}$$

where

$$B = \frac{\ell_v g}{R_v T^2 c_p} - \frac{g}{RT} = \frac{1}{H_v} - \frac{1}{H} \tag{4.32}$$

in which we have used the definitions of the scale heights of total density and water vapour H and H_v respectively (see section 3.2.7), while

$$C = 4\pi\rho_1 NA\left(\frac{R_v T}{e_s(T)} + \frac{\epsilon\ell_v^2}{pTc_p}\right). \tag{4.33}$$

Equations (4.21) and (4.31) can be solved simultaneously to yield the co-evolution of the supersaturation and droplet radii. Figure 4.4 presents a numerically computed example solution which employs some rather drastic simplifying assumptions but adequately illustrates the main qualitative behaviour and orders of magnitude involved. In the early stages, when droplet radii are small, the damping term in (4.31) is weak and s undergoes rapid linear growth. As the particles grow, the damping term eventually overwhelms the source term so that s reaches a maximum and begins to decrease. Decreasing s and increasing particle size makes further growth very slow and the supersaturation also settles on a slowly varying value.

Total droplet number is fixed from the outset in this simplified calculation but in reality new droplets are being continuously nucleated until s stops growing, so the final droplet concentration is decided by the initial distribution of aerosol sizes and the maximum supersaturation reached, about 0.5% in this case. This low value is due to the large number of droplets, due in turn to the large number of condensation nuclei—note that the damping term is proportional to the droplet concentration, so if we lived in a very clean atmosphere where aerosols are two orders of magnitude less numerous than observed, supersaturations of several hundred percent would regularly occur and homogeneous nucleation would rival heterogeneous nucleation.

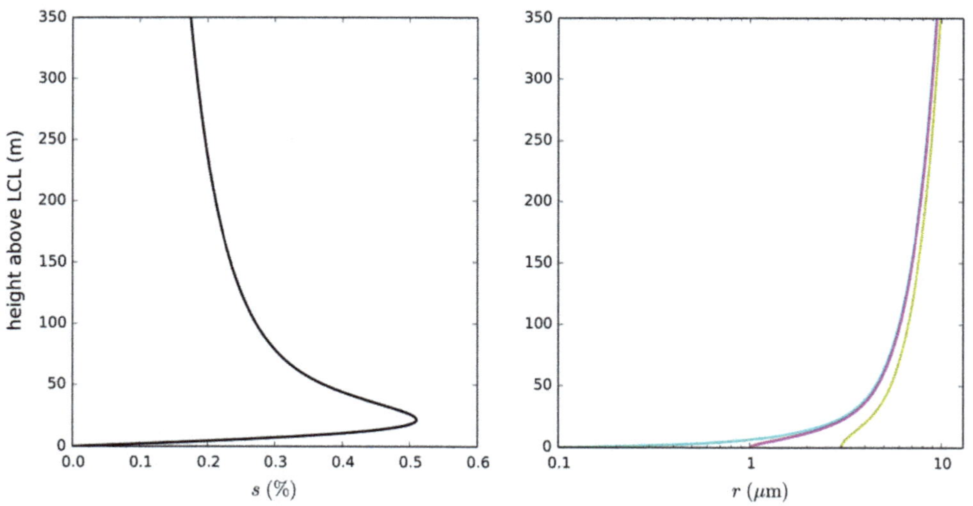

Figure 4.4. Evolution of the supersaturation (left) and droplet radii (right) in an adiabatically lifted parcel, obtained by solving (4.21) and (4.31) assuming constant coefficients $A = 50\ \mu m^2\ s^{-1}$, $B = 5 \times 10^{-4}\ m^{-1}$ and $C = 1/30\ \mu m^{-1}\ s^{-1}$, uniform lifting speed $v_z = 1\ m\ s^{-1}$ and constant total droplet number equally partitioned into three size classes with initial radii $r = 0.1$, 1 and 10 μm, respectively.

A final important point illustrated by this example is the dramatic narrowing of the droplet size spectrum as the cloud evolves: droplet sizes initially span an order of magnitude, but end up closely clustered around 10 μm.

4.4 From cloud droplets to precipitation

4.4.1 The bottleneck

As noted in section 4.2, clouds are colloids—a dispersion of particles suspended in a fluid. The formation of rain, snow, hail and other forms of precipitation collectively known as *hydrometeors* is an example of *colloidal instability*, whereby the suspended particles aggregate into larger units that quickly settle out—precipitate—under gravity (another everyday example is the curdling of milk, though in this case the aggregates are less dense than the surrounding fluid and so they float to the surface rather than fall to the bottom).

Typical convective clouds contain ~10^9 droplets per m^3 with a typical radius of 10 μm and are often observed to precipitate within ~20 min of formation. The resulting shaft of precipitation typically contains around 10^3 drops per m^3, each around 1 mm in size; the 10^6-fold reduction in number is matched by the 10^6-fold increase in particle volume, implying that clouds typically precipitate a large fraction of their total condensate. The precipitation process is thus both efficient and fast, as compared with the time scale for condensational growth discussed in the previous section. On the other hand, some clouds—such as stratus and fair-weather cumulus—do not precipitate at all. Understanding the nature of the precipitation process and why it is triggered in some clouds and not others is a major and, in many respects, unsolved problem in atmospheric physics.

In order for cloud droplets to coalesce into larger aggregates, they must first collide, which requires them to be moving with differential velocities. Cloud drops are in continuous motion due to stirring by small-scale turbulent air currents and by settling under gravity. The viscous drag force is crucial to both types of motion. For spherical cloud droplets of radius less than around 40 μm, the drag force is approximately given by *Stokes' law*

$$F_d = 6\pi\mu r v \tag{4.34}$$

where $\mu = 1.8 \times 10^{-5}$ Pa s^{-1} is the dynamic viscosity of air, r is the droplet radius and v is the droplet's velocity relative to the surrounding air. Thus, a larger droplet will accelerate faster than a smaller one embedded in the same current. Under the action of gravity, the droplets achieve a steady-state settling speed in which the drag force matches the gravitational acceleration g; this so-called *terminal velocity* is given by

$$v_t = \frac{2}{9}\frac{\rho_l g}{\mu} r^2 \tag{4.35}$$

where ρ_l is the density of liquid water and ranges from 1 mm s^{-1} for 3 μm droplets to 10 cm s^{-1} for 30 μm droplets.

Differential droplet velocities clearly increase very rapidly with increasing size difference. However, even particles which are geometrically on a collision course do not always collide: the smaller, slower particle may be pushed aside by the airflow surrounding the larger oncoming particle (called the *collector particle*) and avoid collision. The fraction of potential collisions which lead to actual collisions is called the *collision efficiency*. On the 10–20 min time scales relevant for the formation of precipitation, condensation produces droplets with a typical radius of about 10 μm and detailed calculations show that collision efficiencies in excess of 0.1 require collector droplets with a radius of >20 μm.

Once a non-negligible collision rate is attained, a runaway process ensues: collisions produce larger particles, increasing both the differential velocity and the collision efficiency and making it more likely for further collisions to occur. Collision–coalescence thus provides a very rapid mechanism for the formation of precipitation. The problem, of course, is that cloud droplets initially grow through condensation (collision–coalescence is negligible when particle sizes are below 10 μm) and as we saw in the previous section condensation inherently produces droplet populations with a very narrow size distribution. Condensational growth thus sets up a natural bottleneck for the formation of precipitation; the challenge is to find a robust mechanism that will break this bottleneck by allowing a fraction of the droplets to grow substantially larger than 20 μm and trigger the collision–coalescence process (note that the qualifier 'a fraction' is important: if all the droplets grew to 20 μm, there would be no differential velocities and no collisions).

4.4.2 Precipitation in cold clouds

In *cold clouds*, defined as those whose tops are above the freezing line (i.e. at temperatures less than 0 °C), there is a relatively straightforward way to break the condensational bottleneck. This is called the *Bergeron–Findeisen mechanism*, after the meteorologists who developed the idea in the early 20th century.

When a cloudy air parcel is lifted above the freezing line, the result is a mixture of ice particles and supercooled water (called a *mixed phase* cloud). Not all the droplets immediately freeze because pure water (or a dilute solution) requires temperatures well below 0 °C to freeze spontaneously. At temperatures close to 0 °C, freezing is a heterogeneous process requiring a freezing nucleus (normally called an *ice nucleus* or IN). Only a small fraction of the aerosol population can act as ice nuclei; these are mostly dust aerosols, essentially tiny rock fragments whose crystalline structure has an affinity for that of ice. Some biologically produced compounds and even whole viruses and bacteria can also act as INs. The important point is that INs are invariably in much smaller concentration than CCNs, so that only a small fraction of cloud droplets will freeze.

When a droplet freezes, the result is dramatic: as we saw in section 3.2.6, the saturation vapour pressure over ice is less than over liquid water; at −10 °C the difference is around 10%. As soon as the droplet freezes, the supersaturation it experiences jumps from under 1% to over 10%, and by (4.21) the condensational growth rate increases by an order of magnitude. This allows the frozen particles to quickly increase their size to values well above 20 μm.

To illustrate this point, figure 4.5 shows a highly simplified calculation of condensational growth in a mixed-phase cloud where a fraction f_i of droplets is

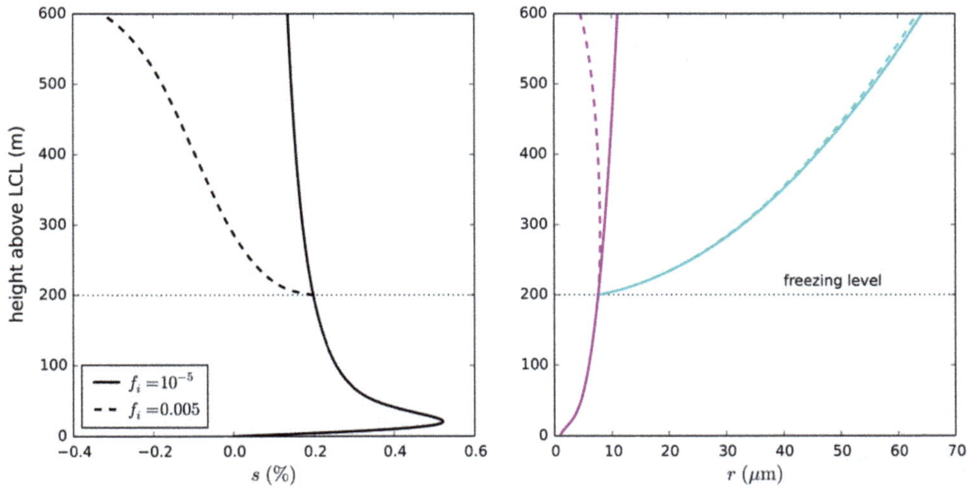

Figure 4.5. Evolution of the supersaturation (left) and droplet radii (right) in an adiabatically lifted parcel in which a fraction f_i of the droplets are assumed to freeze above 200 m. The calculation is identical to that in figure 4.4 except that all droplets have the same initial radius $r = 1$ μm and for the frozen droplets the supersaturation is increased by 10% (i.e. for the frozen droplets, s is the value reported in the left panel plus 10%).

assumed to freeze. The only modification from the calculation shown in figure 4.4 is that the value of s in (4.21) is increased by 10% for the frozen particles[2], while s in (4.31) is computed as an average over the liquid and frozen values weighted by f_i. Moisture will be rapidly depleted around the frozen particles; this may cause the supersaturation to become locally negative, so that supercooled droplets evaporate keeping supersaturation close to zero. After a while, the surrounding liquid droplets will evaporate altogether and the frozen droplet will create a 'hole' where the supersaturation is zero with respect to ice. Growth will then halt unless larger-scale processes keep the humidity and droplet distribution well mixed. We again assume that this mixing happens very efficiently, so that a single value of s is adequate to describe the whole process. If the fraction of frozen droplets is very small, as in the $f_i = 10^{-5}$ case in figure 4.5, this well-mixed value of supersaturation will still be positive; but if the frozen fraction is high enough then the mean supersaturation can drop below zero leading to a drop in the mean size of the liquid droplets (dashed lines in figure 4.5). In the former case the frozen droplets only consume supersaturation made available by lifting, while in the latter case they cannibalize the liquid droplets. In any event, the outcome is essentially identical: the frozen droplets grow within ~10 min to sizes well in excess of 20 μm, where they can initiate the collision–coalescence process.

[2] We emphasize again that this is a crude calculation intended only to illustrate the phenomenon qualitatively. A quantitatively accurate calculation would need to deal with complexities such as the non-spherical shape of ice crystals and the different accommodation coefficients of liquid water and ice (the accommodation coefficient can be defined as the ratio of water molecules striking the droplet surface which are in fact retained within the droplet). More detailed discussion may be found in, e.g. Pruppacher & Klett (2010).

4.4.3 Precipitation in warm clouds

For *warm clouds*—those with a temperature above freezing everywhere—there is no universally agreed-upon solution to the condensational bottleneck problem. Some authors point to the role of *giant condensation nuclei*: observations show that a small fraction of the CCN population is so large that the droplets they nucleate will remain significantly larger than the rest despite condensational narrowing, potentially seeding the collision–coalescence process. Others highlight the role of atmospheric turbulence. Turbulence is highly patchy and strong localized vortices may expel droplets to their boundaries by inertia, leading to a locally enhanced droplet concentration that can increase encounter rates and potentially broaden the droplet size spectrum. Precipitation in warm clouds remains very much an open question in atmospheric science.

Bibliography

Pruppacher H and Klett J 2010 *Microphysics of Clouds and Precipitation* 2nd edn (New York: Springer)

IOP Publishing

Physics of the Atmosphere

Rodrigo Caballero

Chapter 5

Atmospheric radiation

5.1 Electromagnetic radiation

5.1.1 Waves

When a stone is thrown into a pond, it injects KE into the water at the point of impact. This energy is subsequently carried away from the point of impact by wave motion in the water (called *surface or external gravity waves*). A cork floating in the water some distance away will bob up and down (i.e. gain KE) when the waves pass it; we can speak of the waves transporting energy from the point of impact to the cork, which then absorbs some of the energy. Analogously, when a charged object moves in space, it excites *electromagnetic waves*, which carry energy away that may be absorbed by distant objects. Unlike surface gravity waves, which are carried by a material medium, electromagnetic waves are carried by an immaterial medium called the *electromagnetic field*. The immateriality means that electromagnetic waves are free to propagate through a vacuum. They can also propagate through a material medium, though in general they will interact with it, being refracted, absorbed or both.

The classical theory for electromagnetic waves is encapsulated in *Maxwell's equations*, which show that all electromagnetic waves in a vacuum propagate at the same speed $c = 3.00 \times 10^{-8}$ m s^{-1}, the speed of light. The wavelength λ (the distance between two successive crests) and frequency ν (the inverse of the time taken for two crests to pass a given point) are therefore connected by

$$\nu = \frac{c}{\lambda}. \tag{5.1}$$

An equation such as (5.1), linking the wavelength and frequency of a wave, is known as a *dispersion relation*. The relation can also be expressed as

$$\nu = kc \tag{5.2}$$

where $k = 1/\lambda$ is the *wavenumber*, often measured in units of cm^{-1}.

doi:10.1088/978-0-7503-1052-9ch5

The energy carried by wave a depends on the *amplitude*, the deviation of the field from its undisturbed value. Electromagnetic waves are *linear*, meaning that when two waves meet, the total amplitude at any point is simply the sum of the individual amplitudes. This is equivalent to saying that the waves do not interact.

The range of all possible wavelengths, known as the *electromagnetic spectrum*, is conventionally subdivided into a number of sub-ranges. This subdivision is anthropomorphic, the classes corresponding to particular engineering applications or to the way humans perceive the wavelengths in question. At the long-wavelength, low-frequency end ($\lambda > 0.1$ m) we have radio waves, which are used in telecommunications. Proceeding towards shorter waves, we have microwave radiation ($\lambda \approx 1$ mm–10 cm, used in, yes, microwave ovens); infrared radiation ($\lambda \approx 1$ μm–1 mm), so called because it comes before the red end of the visible range and also called thermal radiation, because humans perceive this range as warmth on the skin; visible radiation ($\lambda \approx 0.3$–1 μm), because humans perceive it with their eyes; the ultraviolet, lying beyond the violet end of the visible ($\lambda \approx 10^{-9}$–10^{-7} m); x-rays; and finally gamma rays ($\lambda < 10^{-11}$ m), produced by very energetic nuclear processes.

5.1.2 Photons

Classical wave theory is not sufficient to fully describe radiation in the atmosphere: radiation is the only part of atmospheric science where quantum effects are of the essence. In a quantum description, electromagnetic radiation is considered not as a collection of waves, but as a gas of particles. This wave–particle duality is every bit as weird as it sounds, and is not something we can really 'understand'; rather, it is an empirical fact that we must accept and learn to live with.

The particles, or *quanta*, of electromagnetic radiation are called *photons*. Like any self-respecting particle, a photon carries a discrete amount of energy and linear momentum. However, even when behaving as a particle, radiation somehow 'remembers' its wave nature, and the energy and momentum of a photon are both proportional to its frequency. The energy is

$$E = h\nu \tag{5.3}$$

where $h = 6.626068 \times 10^{-34}$ m^2 kg s^{-1} is *Planck's constant* and the momentum is $h\nu/c$. There is no concept of an amplitude for a photon: if you need more energy at a given wavelength, you just increase the number of photons.

5.2 Interaction between radiation and matter: some generalities

Molecules can interact with radiation in three ways:

1. *Absorption*, in which a photon and a molecule collide, the photon ceases to exist, and its energy is stored as internal energy (rotational, vibrational or electronic) of the molecule. The excited molecule subsequently collides with neighbouring molecules. After a few collisions, the energy absorbed from the photon will be *thermalized*, i.e. randomly distributed among the gas's degrees of freedom in accordance with equipartition. This includes translational degrees of freedom, so the result is a net temperature increase.

2. *Emission*, the opposite of absorption, whereby a molecule whose internal degrees of freedom have been excited through collision spontaneously emits a photon. If the photon then leaves the gas without being absorbed, the result is a net cooling of the gas.
3. *Scattering*, in which a photon and a molecule 'bounce off' each other like billiard balls. A form of scattering occurs when a photon is absorbed and immediately re-emitted without having time to thermalize.

Most of the radiant energy in the atmosphere is contained in visible and infrared photons, with wavelengths spanning the range 0.1–100 μm, energies in the range 10^{-18}–10^{-21} J and linear momenta in the range 10^{-27}–10^{-30} kg m s^{-1}. By comparison, atmospheric molecules have a typical KE ~ 10^{-22} J and momentum ~10^{-24} kg m s^{-1}. Because photon momenta are so much smaller than molecular momenta, scattering has very little effect on molecular speeds; the main effect of scattering is to randomly change the *direction* of photons, and thus to turn a parallel beam into a diffuse photon gas.

On the other hand, absorption can have a large effect on a molecule's total energy. However, to efficiently heat or cool a gas, radiation must be effective in exciting internal degrees of freedom. Monatomic gases, which have no internal degrees of freedom other than electronic, are difficult to heat and cool radiatively. Even if the gas's molecules are polyatomic, the rotation and vibration modes can only couple to the radiation field if the molecule has an *electric or magnetic dipole*. The presence of positive and negative 'poles' (either electric or magnetic) allows radiation to 'latch on' to the molecule and set it spinning or vibrating.

In order for absorption to actually heat a gas, collisions must occur before an absorbed photon has time to be re-emitted. An excited molecular state has a 'half life' for decay, i.e. a typical time after which there is a 50% chance that a photon will be spontaneously emitted. If the half-life is τ and the average time between collisions is τ_c, then efficient thermalization requires $\tau/\tau_c \gg 1$. This condition is true throughout most of the atmosphere.

5.3 Molecular absorption

Quantum mechanics plays an essential role in the absorption of photons by molecules. This section studies the physics of the process in some more detail.

5.3.1 Rotation

In classical mechanics, a particle of mass m moving in a straight line at speed v has momentum mv and KE $mv^2/2$. Analogously, a particle of mass m rotating around an axis with angular velocity ω has *angular momentum*

$$L = I\omega \tag{5.4}$$

and *rotational KE*

$$E_r = \frac{1}{2}I\omega^2 = \frac{L^2}{2I}, \tag{5.5}$$

where

$$I = mr^2 \tag{5.6}$$

is the *moment of inertia* and r is the distance between particle and axis. The moment of inertia expresses the resistance of the particle to rotation; the greater the mass and the greater the distance from the axis, the greater the resistance.

An extended body such as a molecule actually has three moments of inertia, denoted I_1, I_2 and I_3, corresponding to rotation about three orthogonal axes passing through the centre of mass. For a linear molecule like O_2, two of these moments are identical and the third is 0, while an asymmetric molecule like H_2O has three distinct, non-zero moments. For an isolated atom, all three moments are 0.

Rotation is a *bound* motion, meaning it occurs in a confined space, and in quantum mechanics such motions are quantized (linear motion, which is unbound, is not quantized even in quantum mechanics). Angular momentum can only take the discreet values

$$L = \frac{h}{2\pi}\sqrt{j(j+1)} \quad j = 0, 1, 2, \ldots \tag{5.7}$$

corresponding to the energies

$$E_{\rm r} = \frac{h^2}{8\pi I} j(j+1) \quad j = 0, 1, 2, \ldots \tag{5.8}$$

where j is the *quantum number* and I is any of the three moments of inertia of the molecule.

Quantum mechanics also imposes restrictions on the transitions from one energy level to another. The only transitions allowed are those obeying the *selection rule* $\Delta j = \pm 1$, which implies that in a single absorption event, rotational energy can only increase by

$$\Delta E_{\rm r} = \frac{h^2}{4\pi I}(j+1) \tag{5.9}$$

where j is the quantum number of the initial state. Thus, only photons with frequencies

$$\nu = \frac{h}{4\pi I}(j+1) \tag{5.10}$$

can be absorbed.

Exercise 5.3.1: Compute the typical wavelength of photons associated with rotational transitions, given that atmospheric molecules have $m \sim 10$ AMU $= 10^{-26}$ kg and $r \sim 10$ Å $= 10^{-9}$ m, and that the typical atmospheric temperature is 300 K.

5.3.2 Vibration

A reasonable model for a molecule is a number of point masses connected by springs. Consider first a diatomic molecule, which has a single spring. According to classical mechanics, if the two atoms are pulled apart by a distance Δx and released, they will oscillate at the natural (or resonant) frequency

$$\nu_0 = \frac{1}{2\pi}\sqrt{\frac{K}{m}}, \tag{5.11}$$

where K is the spring constant, and the total energy of the motion will be $E = K\Delta x^2/2$.

In quantum mechanics, this bound motion is quantized so that the energy can only take the values

$$E_n = h\nu_0(n + 1/2) \qquad n = 0, 1, 2, 3, \dots \tag{5.12}$$

with the selection rule $\Delta n = \pm 1$. Thus, in an absorption event,

$$\Delta E = h\nu_0 \tag{5.13}$$

which means that only photons with frequency $\nu = \nu_0$ can be absorbed. This is a nice intuitive result: a quantum harmonic oscillator can only absorb photons at its resonant frequency.

For more complicated polyatomic molecules, there will be correspondingly more independent modes of oscillation. In general, a molecule with N atoms will have $3N - 5$ modes if the atoms are arranged in a straight line and $3N - 6$ modes if the atoms are not in a line. Each mode will generally have a distinct resonant frequency and will be associated with a distinct absorption frequency or wavelength. For atmospheric molecules, these are in the range 1–20 μm.

5.3.3 Roto-vibrational transitions

Nothing prevents vibrational and rotational transitions from occurring simultaneously; these are called *roto-vibrational transitions*. The result is a cluster of absorption lines grouped around the central vibration-only line.

5.3.4 Line broadening

A useful way to characterize absorption is through the *mass absorption coefficient* $k_a(\nu)$, which is proportional to the probability that a photon of frequency ν will be absorbed when colliding with a molecule (a more precise definition will be given in section 5.6.3 below). Until now, we have been assuming that the only photons absorbed are those with *exactly* the energy (or frequency) required for an allowed transition, so we can write

$$k_a = S\,\delta(\nu - \nu_0) \tag{5.14}$$

where ν is the frequency of the incident photon, ν_0 is the frequency at which absorption can occur, S is called the *line strength* and δ is the Dirac delta function. A plot of this function looks like a vertical line at $\nu = \nu_0$, hence the term *absorption line*.

In reality, absorption lines are not so sharp: they are *broadened* by various physical processes. As a result,

$$k_a = S\,f(\nu - \nu_0) \tag{5.15}$$

with

$$\int_0^\infty f(\nu - \nu_0) = 1, \tag{5.16}$$

where f is a *line shape* of finite width whose precise form depends on the underlying broadening process.

One process is *Doppler broadening*. An incoming photon of frequency ν_0 (as measured by a stationary observer) will be 'seen' by a molecule moving at speed u to have a Doppler-shifted frequency $\nu = \nu_0 + u/c$ (assuming $u \ll c$). Thus, a photon that is slightly de-tuned with respect to the resonant frequency still has a chance of being absorbed, if the molecule it encounters is moving at the right speed. The distribution of speeds follows the Maxwell–Boltzmann distribution (section 2.1.3), and it can be shown that this results in a Gaussian line shape

$$f = \frac{1}{\gamma_D \sqrt{\pi}} e^{-\left(\frac{\nu-\nu_0}{\gamma_D}\right)^2} \tag{5.17}$$

where

$$\gamma_D = \frac{\nu_0}{c}\sqrt{\frac{2kT}{m}} \tag{5.18}$$

is the half-width at half height. Thus, the line will broaden as the gas warms.

A second process is *pressure (or collisional) broadening*. The issue here is that collisions generally jolt a molecule out of whatever roto-vibrational state it happens to find itself in. Thus the average time spent by a molecule in any given state is τ_c, the mean time between collisions. By Heisenberg's principle, this implies an uncertainty in the state's energy of

$$\Delta E \sim \frac{h}{\tau_c}. \tag{5.19}$$

There will be a similar uncertainty in the energy *difference* between states, implying an uncertainty in the frequency of absorbed photons

$$\Delta\nu \sim \frac{1}{\tau_c}. \tag{5.20}$$

A more detailed calculation shows that the actual line shape in this case is Lorentzian:

$$f = \frac{\gamma_L}{\pi}\frac{1}{(\nu - \nu_0)^2 + \gamma_L^2}, \tag{5.21}$$

with a width

$$\gamma_L = \frac{1}{2\pi\,\tau_c}. \tag{5.22}$$

The collision time τ_c can be related to the temperature and pressure of the gas using kinetic theory. The collision time is given by

$$\tau_c \langle v \rangle \sigma n = 1, \tag{5.23}$$

where $\langle v \rangle$ is the mean speed of the molecules, σ is (roughly) the mean cross-section of a molecule and n is the number density. Using $p = nkT$ and $kT = m\langle v^2 \rangle/3$, we then have

$$\gamma_L \approx \frac{\sigma}{2\pi}\sqrt{\frac{3kT}{m}}\frac{p}{kT}. \tag{5.24}$$

Thus the line will broaden proportionally to the pressure.

Exercise 5.3.4: Explain why (5.23) is true, and derive (5.24).

Pressure broadening dominates in the lower atmosphere, where pressure is high, while Doppler broadening dominates in the upper atmosphere. The crossover occurs where

$$\frac{\gamma_L}{\gamma_D} \sim \sigma\lambda_0 n \sim \frac{\sigma\lambda_0}{m}\rho_0 e^{-z/H} = 1 \tag{5.25}$$

where $\lambda_0 = c/\nu_0$, ρ_0 is the surface air density and H is the scale height. Taking $\sigma = 10^{-19}\,\mathrm{m}^2$, $\lambda_0 = 10\,\mu\mathrm{m}$, $m = 30$ AMU and $\rho_0 = 1\,\mathrm{kg\,m}^{-3}$, we find

$$z \sim 3H, \tag{5.26}$$

so pressure broadening dominates over Doppler broadening throughout the bulk of the atmosphere.

5.4 Scattering

We saw above that absorption is characterized by the single quantity k_a, the absorption coefficient. Two quantities are required to characterize scattering: the *mass scattering coefficient* k_s, proportional to the probability that a photon will be scattered when meeting a particle (see section 5.6.3 for a more precise definition) and the *scattering phase function* $P(\theta)$, which is the probability that the outgoing photon will be directed at an angle θ relative to the direction of incidence.

Unlike molecular absorption, scattering in the atmosphere (including scattering by molecules) can be adequately treated using only classical electromagnetism. In the early 20th century, the German physicist Gustav Mie worked out a complete solution to Maxwell's equations for electromagnetic waves interacting with a homogeneous sphere. Cloud droplets and small raindrops are essentially spheres, so Mie's solution applies accurately. The theory also works well for molecules—somewhat surprisingly, since molecules are neither homogeneous nor spherical ('works well' means that the theory's predictions agree well with experiment). On the other hand, the theory works less well for ice particles and aerosols, which can have very irregular shapes; there is no complete theory for scattering by such particles.

Mie's solution gives explicit formulae for $P(\theta)$ and for the *scattering efficiency* Q_s, an adimensionalized scattering coefficient:

$$Q_s = \frac{m}{\pi r^2} k_s, \tag{5.27}$$

where m is the mass of the scattering particle and r its radius. The solutions depend on only two adimensional parameters: the *scattering parameter*

$$x = 2\pi \frac{r}{\lambda}, \tag{5.28}$$

where λ the wavelength, and the the *relative refractive index*

$$N_r = \frac{N_p}{N_s}, \tag{5.29}$$

where N_p is the refractive index of the particle and N_s that of the surrounding medium (recall that the refractive index measures the deviation of a light ray when passing from one medium to another).

In the rest of this section, we discuss some features of Mie's solutions (we will not look at the mathematical theory, which is very complicated and not very enlightening). A summary of the solutions' behaviour is shown in figure 5.1 and a plot of Q_s as a function of x is shown in figure 5.2. A number of different scattering regimes arise. At very small x, Mie's theory predicts vanishing k_s. In this regime, which is relevant to infrared radiation interacting with molecules, there can be absorption but no scattering. At larger x we have *Rayleigh scattering*, then full *Mie scattering* and finally *geometric optics*. We discuss each of these below.

5.4.1 Rayleigh scattering

For x in the range 0.002–0.2 we are in the *Rayleigh scattering regime*. The name comes from an earlier theory by Lord Rayleigh, which is a limiting case of the more general Mie theory. In this regime, which is relevant to visible radiation interacting with molecules and for infrared radiation interacting with aerosols,

$$k_s \sim \frac{r^3}{\lambda^4}, \tag{5.30}$$

which implies that short wavelength (blue) light is scattered more strongly than long wavelength (red) light. This wavelength dependence explains the blue colour of the clear sky (when viewed at an angle away from the Sun) and the red colour of sunsets (see also below). The phase function has the form

$$P(\theta) = \frac{3}{4}\left(1 + \cos^2\theta\right), \tag{5.31}$$

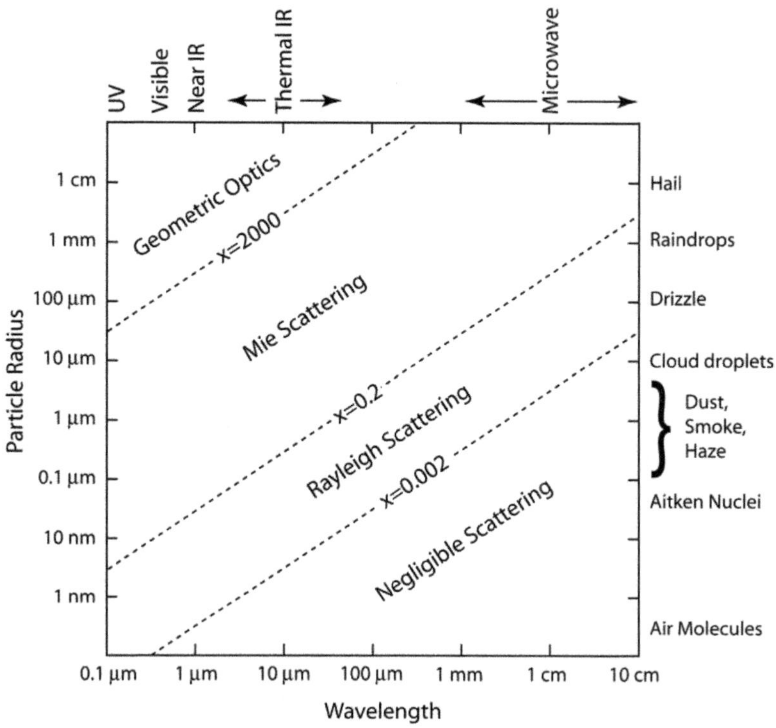

Figure 5.1. Regimes of atmospheric scattering. From Petty (2004).

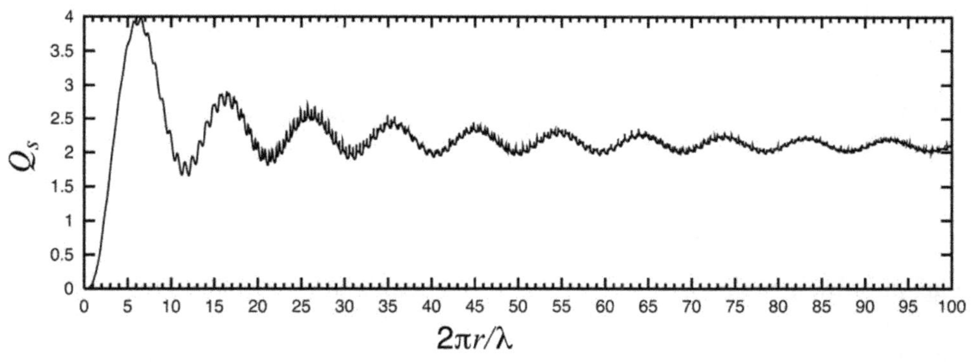

Figure 5.2. Scattering efficiency $Q_s = m(\pi r^2)^{-1}k_s$ for a sphere of refractive index 1.33 in air. From Petty (2004).

a plot of which is shown for $x = 0.1$ in figure 5.3. The function has mirror symmetry around the midpoint, which means that a photon has equal chances of being scattered forward or backward.

5.4.2 Mie scattering

As x increases beyond ~0.2, Rayleigh theory becomes inaccurate and the full Mie theory comes into force. At first, Q_s continues to increase, reaching a very high peak

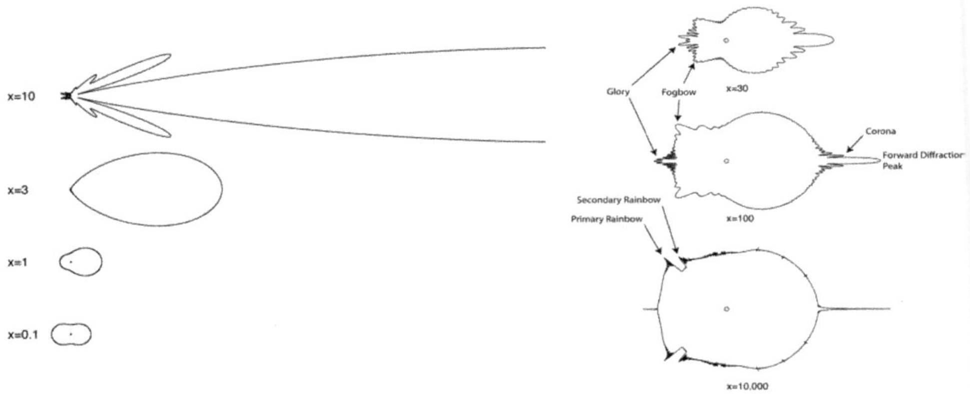

Figure 5.3. Polar plot of the scattering phase function $P(\theta)$ for various values of the scattering parameter x. In all cases the incoming photon is moving in horizontally from the left. The central dot represents the scattering particle. From Petty (2004).

somewhere in the region $x = 1$–10. Particles in this regime have a very strong reddening effect and largely explain the very intense red observed in sunsets. Beyond the first peak, Q_s drops down steeply. This leads to *blueing*: if all atmospheric particles fell into this regime, sunsets would look blue instead of red. Blueing does occur naturally, but it is very rare ('once in a blue moon').

At even greater values of x, the scattering efficiency settles around a value of 2 (see figure 5.2). In this regime, relevant to both visible and infrared radiation interacting with cloud droplets, the scattering efficiency does not change much with λ, which is why clouds look white. Though the scattering efficiency is very high for clouds, the phase function (figure 5.3) is very asymmetric, with a very strong forward peak. Thus, even on an overcast day, plenty of sunlight still makes it to the ground. However, the light will have been scattered and is *diffuse*—it casts no shadows.

5.4.3 Geometric optics

At very high x, Mie theory converges with *geometric optics*, the macroscopic viewpoint where light rays travel in straight lines and are refracted (bent) when the index of refraction changes. This regime is applicable to visible radiation interacting with very large cloud droplets and rain. It is here that we find the most spectacular phenomena of atmospheric optics: rainbows, fogbows, coronas, halos, glories, etc. From the geometric optics point of view, rainbows can be understood by the refraction and internal reflection of light rays in spherical drops. The rainbow is seen at the critical angle where the light rays bunch up and the spectral splitting is due to the fact that the critical angle depends on the wavelength—it is less for longer wavelength, so red ends up as the outer circle of the rainbow.

From the Mie theory point of view, rainbows arise as a peak in the scattering phase function (see figure 5.3). The rather peaky structure of the phase function at high x also allows us to rationalize other notable phenomena. *Coronas* and *halos* are the coloured rings seen around bright objects viewed through a thin cloud (figure 5.4(a)); they are due to secondary peaks near the main forward peak.

Figure 5.4. (a) A full moon seen through clouds, surrounded by a corona or halo. (b) A fogbow (large white bow) with a glory at the centre. The Sun is behind the observer. (c) A solar glory viewed from an aeroplane.

Fogbows are similar to rainbows but caused by fog or cloud droplets; the peak involved is quite broad (figure 5.3), so a sharp spectral splitting is not obtained and fogbows look whitish (figure 5.4(b)). Finally, *glories* are coloured circles seen around a shadow projected onto a cloud (figure 5.4(c)); they are due to sharp peaks very close to the backward scattering peak.

5.5 Absorption by cloud drops

In general, all particles are capable of absorbing as well as scattering. In addition to Q_s, Mie's solution also gives the *absorption efficiency* Q_a, related to the absorption coefficient by

$$Q_a = \frac{m}{\pi r^2} k_a. \tag{5.32}$$

For molecules, Mie's solution gives the wrong result since quantum mechanics is essential for molecular absorption (section 5.3), but it works well for macroscopic particles such as cloud droplets. The role of absorption relative to scattering is measured by the *single scatter albedo*

$$\omega = \frac{k_s}{k_a + k_s}, \tag{5.33}$$

which is the probability that a photon will be scattered rather than reflected upon interacting with a particle—it can be thought of as the reflectivity of a single particle.

The single scatter albedo for a range of typical cloud drop radii is shown in figure 5.5. The main point to take from this figure is that cloud drops scatter without absorption in the visible range (where $\omega = 1$), while absorption is very strong in the infrared: pure water is transparent to visible but opaque to infrared radiation. As a result, clouds are very efficient scatterers of solar radiation and very efficient absorbers of infrared.

5.6 Radiative transfer: generalities

5.6.1 Energy flux in a gas of photons

Having dealt with the interaction of a single photon with a single particle, the rest of this chapter studies the interaction between a gas of photons and a collection of

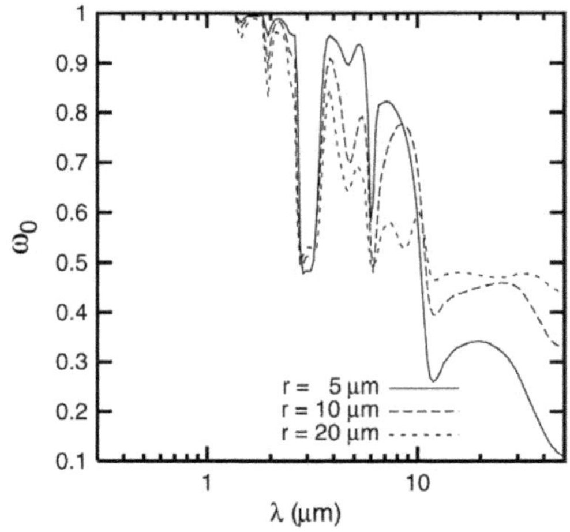

Figure 5.5. Single scatter albedo ω for water droplets of radius r over the visible and infrared ranges. From Petty (2004).

atmospheric particles. We are particularly interested in the energy transported by radiation. The convergence of the energy flux determines the radiative heating/cooling rate, which links radiation to atmospheric thermodynamics.

We begin in this section by defining some measures of the energy flux carried by a photon gas.

Radiance or intensity is the energy crossing a unit area with normal **n** in unit time, carried by photons of wavelength λ travelling parallel to **n**; it has units of W m^{-2} m^{-1} sr^{-1}. We will denote intensity with the symbol $I(\lambda, \mathbf{n})$.

Irradiance or flux is the energy crossing a unit area with normal **n** in unit time, carried by photons of wavelength λ and *summed over all directions of incidence in a hemisphere*:

$$F(\lambda, \mathbf{n}) = \int I(\lambda, \mathbf{n}') \, \mathbf{n} \cdot \mathbf{n}' \mathrm{d}\mathbf{n}' = \int_{\varphi=0}^{2\pi} \int_{\theta=0}^{\pi/2} I(\lambda, \theta, \varphi) \cos\theta \sin\theta \, \mathrm{d}\theta \, \mathrm{d}\varphi \tag{5.34}$$

where $\mathbf{n}'$ is the direction of incidence and θ, φ are, respectively, the zenith and azimuth angles between **n** and $\mathbf{n}'$. Irradiance has units W m^{-2} m^{-1}.

Integrating over wavelength, we obtain the *total flux* $F(\mathbf{n})$ (W m^{-2}), the total energy crossing unit area normal to **n** in unit time:

$$F(\mathbf{n}) = \int_0^\infty F(\lambda, \mathbf{n}) \, \mathrm{d}\lambda. \tag{5.35}$$

Note the distinction between radiance and irradiance: radiance is an intrinsic property of the photon gas, while irradiance is the relevant quantity if we are interested in the flux of energy through a pre-defined surface (e.g. the radiant energy absorbed at the ground).

5.6.2 Black body radiation

Consider a box or cavity containing a molecular gas interacting with a gas of photons. The sides of the box are perfectly reflecting, while the gas itself is perfectly absorbing—it has an infinite number of absorption lines so it can absorb *and emit* photons with any desired wavelength. Assume that the molecular gas is in equilibrium so that it has a Maxwell–Boltzmann distribution. In this case, we also expect the photon gas to have a steady-state radiance. German physicist Max Planck was the first to theoretically derive the form of this radiance:

$$B = \frac{2hc^2}{\lambda^5} \frac{1}{e^{\frac{hc}{k\lambda T}} - 1} \tag{5.36}$$

which is now called the *Planck function*. This equilibrium radiation field is known as *cavity radiation*. Note the many similarities with the Maxwell–Boltzmann distribution: it is isotropic, it depends only on the temperature T of the system and it contains the exponent $hc/\lambda kT$, the ratio of photon energy to mean thermal energy, analogous to the exponent $mv^2/2kT$ in the Maxwell–Boltzmann distribution.

Imagine punching a small hole in one side of the cavity, small enough that any photon *entering* the box from the outside will be unable to find its way out again before being absorbed and thermalized. Such a hole is perfectly black, in the sense that it absorbs all radiation falling upon it. That does not mean it cannot *emit*: some photons from the equilibrium radiation field in the cavity will find their way out of the hole and if we measure their energy distribution we will find it to follow the Planck function.

This thought experiment motivates the definition of a *black body* as any object that is perfectly absorbing and is in thermodynamic equilibrium (i.e. has a uniform, constant temperature). We will see below that *all* black bodies emit radiation following the Planck function. Thus cavity radiation is also known as *black body radiation*.

As shown in figure 5.6, the dependence of the Planck function on temperature is very strong: peak radiance changes by an order of magnitude over the range of temperatures experienced in Earth's atmosphere. Note also that the peak of the function moves to shorter wavelengths as temperature increases. When the temperature reaches ~6000 K (roughly the temperature of the solar surface) the peak is in the visible range. The right panel in figure 5.6 shows that solar and terrestrial radiation occupy two distinct bands with very little overlap, which allows a convenient separation of the atmospheric radiative transfer problem into two parts, one dealing with absorption and scattering of solar or *shortwave* radiation, the other with absorption and emission of terrestrial or *longwave* radiation.

Following (5.34), the irradiance emitted by a black body is

$$\int_{\varphi=0}^{2\pi} \int_{\theta=0}^{\pi/2} B \cos\theta \sin\theta \, d\theta \, d\varphi = \pi B, \tag{5.37}$$

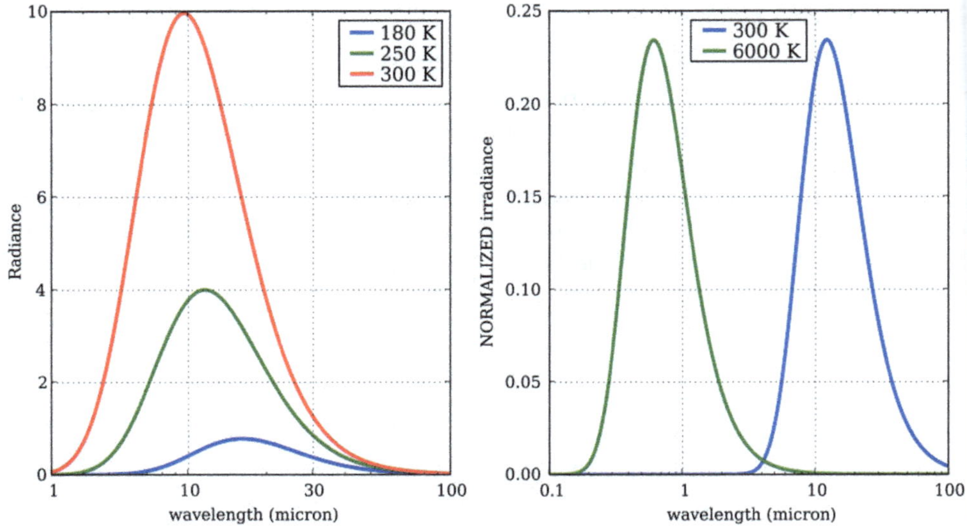

Figure 5.6. (Left) Black body radiance B (W m^{-2} μm^{-1} sr^{-1}) for three typical Earth-like temperatures. (Right) Black body irradiance πB for typical terrestrial and solar temperatures, normalized by $\lambda\sigma T^4$ so that the structure of the two curves can be easily compared.

and if we also integrate over λ we obtain *Stefan's law*, which states that each squared metre of a black body's surface emits radiant energy at the rate

$$F = \sigma T^4 \tag{5.38}$$

where

$$\sigma = \frac{2\pi^5 k^4}{15 c^2 h^3} \tag{5.39}$$

is the *Stefan–Boltzmann* constant, which has the value 5.67×10^{-8} W m^{-2} K^{-4}.

5.6.3 Extinction and optical thickness

Consider a beam of photons all having the same wavelength and direction. Imagine we point the beam perpendicularly at a slab of atmosphere. As the beam passes through the slab, some of the photons will be absorbed and some scattered away from their initial direction. Thus the intensity of the beam will diminish. The general term for attenuation through either absorption, scattering or both together is *extinction.*

If the slab is very thin, then the total amount of extinction (i.e. the difference in intensity between outgoing and incoming beams) turns out to be proportional to the amount of matter in the slab and to the intensity itself:

$$dI = -Ik_a\rho_a \, ds - Ik_s\rho_s \, ds \tag{5.40}$$

where I is the radiance of the incoming beam, dI is the difference between incoming and outgoing radiances, k_a is the *mass absorption coefficient* (see section 5.3.4), ρ_a is the mass density of absorbers, k_s is the *mass scattering coefficient* (section 5.4), ρ_s is the mass density of scatterers and ds is the width of the slab. This linear relationship between extinction, mass of absorbers/scatterers, and intensity is called the *Beer–Lambert law*.

The coefficients k_a and k_s (which have units $m^2\ kg^{-1}$) can be defined as the constants of proportionality in this law. They can also be defined as the *probability that a photon will be absorbed/scattered when it traverses a unit amount of mass which is uniformly distributed in a tube of unit cross section.*

If the same particles are doing the absorbing and scattering, $\rho_a = \rho_s = \rho$ and (5.40) can be written

$$\frac{d \ln I}{ds} = -k_e \rho \tag{5.41}$$

where

$$k_e = k_a + k_s \tag{5.42}$$

is called the *mass extinction coefficient*. Defining the *optical thickness*

$$\tau = \int_0^s k_e \rho \, ds, \tag{5.43}$$

(5.41) has the solution

$$I(s) = I(0)\, e^{-\tau(s)} \tag{5.44}$$

where s measures distance along the path of the beam, with $s = 0$ at the point of entry. Thus, radiance decays exponentially after entering an absorbing/scattering medium.

In general, the extinction coefficient will change with position: for instance, in regions where pressure is higher, collisional line broadening will lead to greater k_a. In those cases where k_e varies weakly over the path, then we can write

$$\tau(s) = k_e u(s), \tag{5.45}$$

where $u(s)$ is the *mass path*—the mass of absorbers in a tube of length s and unit cross section, coaxial with the beam.

5.6.4 Transmissivity and absorptivity

Consider again the beam of radiation incident on a slab of width s, as in the previous section. The *transmittance* or *transmissivity* $\mathcal{T}$ of the slab is defined as the fraction of the incident radiance that makes it through to the other side:

$$\mathcal{T} = \frac{I(s)}{I(0)} = e^{-\tau}. \tag{5.46}$$

The *absorptance* or *absorptivity* $\mathcal{A}$ is the fraction of the incident radiance that is absorbed by the slab. *If there is no scattering* ($k_s = 0$), we have

$$\mathcal{A} = 1 - \mathcal{T} = 1 - e^{-\tau}. \tag{5.47}$$

Note that in order to conserve energy, we must have $0 \leqslant \mathcal{T} \leqslant 1$ and $0 \leqslant \mathcal{A} \leqslant 1$.

By convention, a body is called *optically thick* if it has $\tau > 1$, so that $\mathcal{T} < 1/e$, and *optically thin* if $\tau < 1$.

5.6.5 Emissivity and Kirchhoff's law

Any object with a non-zero temperature emits some amount of radiation. This is because temperature implies an agitation of the molecules making up the object, which inevitably leads to excitation of the electromagnetic field. The question is, how *much* radiation does it emit and how does the emitted radiance depend on temperature? At first sight, it might seem like objects have a great deal of freedom to emit just as much radiation as they like; the amount emitted will depend on the details of what the body is made of. This is not altogether true: as we will see in this section, thermodynamics imposes some surprisingly strong constraints.

The *emittance* or *emissivity* ϵ of a body at temperature T is defined as the ratio of the radiance I emitted by the body to the radiance emitted by a black body at the same temperature:

$$\epsilon = \frac{I}{B(T)}. \tag{5.48}$$

Kirchhoff's law states that *absorptivity and emissivity are always equal*:

$$\epsilon = \mathcal{A}. \tag{5.49}$$

Note that this is valid for each wavelength. Kirchhoff's law means that black body radiance is the *maximum* radiance a body can emit at a given temperature. Also, any object that is perfectly black ($\mathcal{A} = 1$) must emit as a black body, independently of what the body is made of.

Kirchhoff's law is a consequence of the second law of thermodynamics. Consider an object at temperature T_1 exchanging radiation with a black body at temperature T_2. The rate of energy transfer from the object to the black body is $\epsilon B(T_1)$, while the transfer from black body to object is $\mathcal{A}B(T_2)$. The second law implies that this heat exchange must result in the colder body warming at the expense of the warmer body. Once the bodies reach equilibrium, the net flux must be zero, so $\epsilon B(T_1) = \mathcal{A}B(T_2)$, and the temperatures must be equal, so $B(T_1) = B(T_2)$; Kirchhoff's law then follows. The validity at each wavelength can be shown by repeating the argument, but this time inserting a filter between the two bodies which only allows a given wavelength to pass. Closer examination shows this demonstration to have hidden flaws, which means the Kirchhoff's law *can* be violated under certain circumstances—this does not imply a violation of the second law, only a limitation of the validity of Kirchhoff's law. In practice, such violations are rare. Kirchhoff's law may be considered valid for all processes of relevance to the atmosphere.

5.6.6 Equivalent width and line saturation

Often we are interested in the total absorptivity integrated over a certain frequency range—for instance, we may be interested in the total absorptance due to a

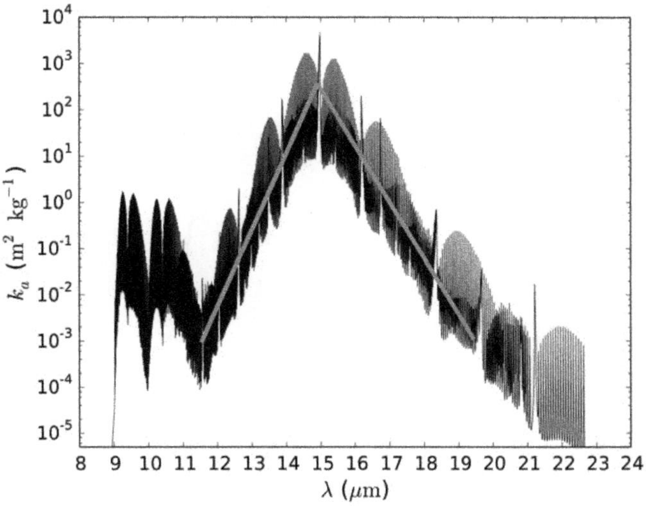

Figure 5.7. Mass absorption coefficient for CO_2 at 500 hPa in the spectral interval containing the 15 μm vibration–rotation band. The thick grey line shows an approximate exponential fit as in (5.51).

particular spectral line, or to a collection of lines grouped into an absorption band. The band-integrated absorptivity is called the *equivalent width*

$$\mathcal{W} = \int_{\nu_1}^{\nu_2} \mathcal{A}\mathrm{d}\nu = \int_{\nu_1}^{\nu_2} \left(1 - \mathrm{e}^{-\tau}\right) \mathrm{d}\nu, \tag{5.50}$$

where τ here is the absorption optical thickness and the integral extends over an interval containing the band in question. $\mathcal{W}$ is the width that an idealized, top hat-shaped absorption band (with 100% absorptance in-band and 0% outside) would need to have in order to match the total absorptivity in the interval considered.

As an example, we will consider the equivalent width of the 15 μm vibration–rotation band of CO_2, shown in figure 5.7. This is a strong band close to the Planck function peak at terrestrial temperatures, making CO_2 an important player in Earth's radiation budget. The band consists of a very large number of individual absorption lines, with line strengths decaying roughly exponentially away from the peak at 15 μm. If we assume that the lines are broadened to the point that they substantially blend into each other, forming a continuous spectrum, we may approximately write

$$k_\mathrm{a}(\nu) \approx k_0\, \mathrm{e}^{-\frac{|\nu-\nu_0|}{\gamma}} \tag{5.51}$$

where k_0 is the peak value of k_a, attained at ν_0 and γ is the rate of decay (a sketch of this fit is shown in figure 5.7). Neglecting any spatial variability in k_a, we then have

$$\mathcal{W} = \int_{\nu_1}^{\nu_2} \left(1 - \mathrm{e}^{-\tau_0 \mathrm{e}^{-|\nu-\nu_0|/\gamma}}\right) \mathrm{d}\nu \tag{5.52}$$

where $\tau_0 = uk_0$, with u the mass path, and ν_1, ν_2 is a suitable interval symmetric about ν_0.

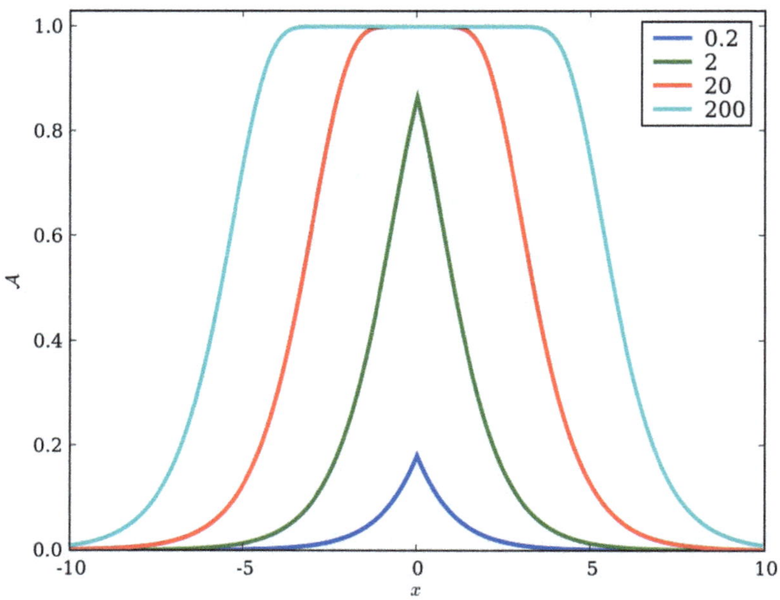

Figure 5.8. The absorptivity $\mathcal{A} = 1 - e^{\tau_0 e^{-|x|}}$ for the values of τ_0 indicated.

This integral cannot be explicitly evaluated, but we can obtain an approximate solution. First, we rewrite (5.52) as

$$\mathcal{W} = 2\gamma \int_0^{x_1} \left(1 - e^{-\tau_0 e^{-|x|}}\right) dx \tag{5.53}$$

where we have set $x = (\nu - \nu_0)/\gamma$. We consider two limiting cases:

Weak line. If $\tau_0 \ll 1$, then we can expand the exponential to first order:

$$\mathcal{W} \approx 2\gamma \int_0^{x_1} \tau_0 e^{-x}\, dx \approx 2\gamma\tau_0, \tag{5.54}$$

assuming $e^{-x_1} \approx 0$. In this limit, $\mathcal{W}$ is *linear* in the mass path u (recall $\tau_0 = uk_0$): doubling the CO_2 concentration will double the total absorptivity of the band.

Strong line. If $\tau_0 \gg 1$, then the absorptivity profile takes on a somewhat top-hat appearance, as can be seen in figure 5.8: the centre of the band is *saturated* ($\mathcal{A} \approx 1$), but further out there is a sharp transition to small $\mathcal{A}$. The 'shoulder' is at $\nu \approx \ln(\tau_0)$. The part of the integral falling outside the saturated centre gives only a small contribution to the total, so

$$\mathcal{W} \approx 2\gamma \int_0^{\ln(\tau_0)} dx = 2\gamma \ln(\tau_0). \tag{5.55}$$

In this limit, $\mathcal{W}$ depends *logarithmically* on mass path. A doubling of CO_2 increases $\mathcal{W}$ by $2\gamma \ln 2$; to obtain the same increase again, we need to quadruple the concentration.

The transition from linear mass path dependence to slower dependence at high u is referred to as *line (or band) saturation.* At the current concentration of CO_2 in the Earth's atmosphere (380 ppm), the 15 μm band is saturated (see exercise below) and the atmospheric greenhouse effect depends only logarithmically on the concentration.

Exercise 5.6.6: Estimate the CO_2 mass path (in kg m^{-2}) encountered by a beam of radiation travelling upwards from the surface to the top of the atmosphere. Using figure 5.7, estimate τ_0.

5.6.7 The Schwarzschild equation

Consider again a parallel beam of radiance I incident perpendicularly on a slab of thickness ds. The radiance exiting perpendicular to the slab will have a contribution from the incident beam (attenuated by absorption and scattering) and in general there will also be a contribution by emission from the slab. Also, photons within the slab may be scattered into the direction of the beam and give an extra contribution to the exiting radiance.

Taking these contributions into account, the Beer–Lambert law (5.40) generalizes to the *Schwarzschild equation*

$$\mathrm{d}I = -Ik_e\rho \mathrm{d}s + Jk_e\rho \mathrm{d}s \tag{5.56}$$

where J is the *source function*. When scattering is involved, the source function is complicated, involving an integral over all directions which can scatter into the outgoing beam.

5.7 Radiative transfer of longwave radiation

Things are simpler if only emission contributes to the source function and not scattering. This is an appropriate approximation for longwave radiation in the Earth's atmosphere, but not for solar radiation. In this case, $k_e = k_a$ and $k_a\rho \mathrm{d}s = \mathrm{d}\tau$ is the absorptivity of the slab. Using Kirchhoff's law, this is also the emissivity of the slab, so $J = B$ (the Planck function) and Schwarzschild's equation becomes

$$\frac{\mathrm{d}I}{\mathrm{d}\tau} = -I + B. \tag{5.57}$$

Multiplying by e^{τ} and integrating over τ, we find

$$I(s) = I(0)e^{-\tau} + \int_0^{\tau} B\, e^{-(\tau-\tau')}\mathrm{d}\tau'. \tag{5.58}$$

The first term is the attenuated contribution from the incident beam; the second term is a sum over contributions emitted by thin layers between 0 and s, each attenuated according to its distance from s. Note that temperature, and therefore B, may vary along the path.

5.7.1 Plane parallel approximation

Given the temperature structure, the Schwarzschild equation gives us the radiance everywhere in the atmosphere. But it is the *irradiance* that determines the energy flux entering or leaving atmospheric parcels and provides the link with thermodynamics and dynamics. Irradiance involves an integral over directions which can be tricky to deal with. To simplify matters, it is common to make the *plane parallel* approximation, which consists in neglecting *horizontal* variations in temperature and density, as if the atmosphere were made up of horizontal slabs each of uniform temperature. This approximation works because of the small aspect ratio of the atmospheric temperature and pressure: the atmosphere has a depth scale of ~10 km, but appreciable horizontal temperature and pressure changes only occur over scales ~100–1000 km. Numerical atmosphere models typically have a horizontal grid spacing ~100 km and temperature is assumed to be uniform within each grid box. For reasons that will be clear later (section 5.7.3), the radiance reaching a certain point P comes mostly from distances less than 1 optical thickness unit away from P (shown by the dashed line in figure 5.9). Earth's atmosphere is fairly opaque in the infrared: the optical thickness over 1 scale height is about 1. Thus, radiation reaching P comes from points no further than about 10 km away, which means horizontal fluctuations can be ignored while vertical variations cannot be ignored. However, the approximation becomes questionable in the presence of clouds, where the absorption and scattering coefficients can change by an order of magnitude over 0.1–1 km.

Taking z as the vertical axis, θ as the zenith angle and $\mu = \cos\theta$ (see figure 5.9), we can write

$$\tau(s) = \tau(z, \mu) = \int_0^z k_a(z)\rho(z)\frac{dz}{\mu} = \frac{\tau(z)}{\mu} \tag{5.59}$$

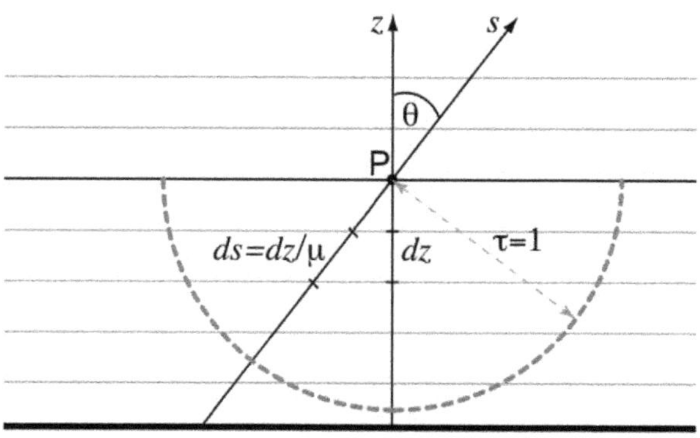

Figure 5.9. An illustration of the plane-parallel approximation in an idealized atmosphere in which absorber concentration is uniform horizontally but varies vertically. The dashed grey line shows a spherical neighbourhood of radius 1 optical thickness unit around an arbitrary point P.

where $\tau(z)$ is the *optical depth*: the optical thickness measured vertically upwards from the surface to z. Furthermore, the solution to the Schwarzschild equation, (5.58), can be written

$$I(z) = B_s e^{-\tau(z)/\mu} + \int_0^z B \frac{d}{dz'} e^{-[\tau(z)-\tau(z')]/\mu} dz' \tag{5.60}$$

where we have assumed that the surface emits black body radiation B_s.

5.7.2 Two-stream approximation

Now consider the irradiance $F^{\uparrow}(z)$ flowing *upward* across some level z:

$$F^{\uparrow}(z) = \int_{\varphi=0}^{2\pi} \int_{\theta=0}^{\pi/2} I(z) \cos\theta \sin\theta \, d\theta \, d\varphi = 2\pi \int_0^1 I(z)\mu d\mu. \tag{5.61}$$

Substituting (5.60) into this expression yields

$$F^{\uparrow} = \pi B_s \langle e^{-\tau/\mu} \rangle + \int_0^z \pi B \frac{d}{dz'} \langle e^{-(\tau-\tau')/\mu} \rangle dz', \tag{5.62}$$

where

$$\langle e^{-\tau/\mu} \rangle = 2 \int_0^1 e^{-\tau/\mu} \mu d\mu. \tag{5.63}$$

is the cosine-weighted average transmissivity. This integral cannot be done analytically. The traditional way forward is to use the approximation introduced by Elsasser (1942):

$$2 \int_0^1 e^{-\tau/\mu} \mu d\mu \approx e^{-1.66\tau} \tag{5.64}$$

where the 1.66 coefficient is called the *diffusivity factor*. As shown in figure 5.10, the approximation is particularly good for $\tau < 1$.

If we re-define the symbol τ to include the diffusivity factor, then the upward irradiance becomes

$$F^{\uparrow} = \pi B_s e^{-\tau} + \int_0^z \pi B \frac{d}{dz'} e^{-(\tau-\tau')} dz', \tag{5.65}$$

and the downward irradiance

$$F^{\downarrow} = \int_z^{\infty} \pi B \frac{d}{dz'} e^{-(\tau'-\tau)} dz'. \tag{5.66}$$

These fluxes are solutions to the *two-stream radiative transfer* equations

$$\frac{dF^{\uparrow}}{d\tau} = -F^{\uparrow} + \pi B \tag{5.67}$$

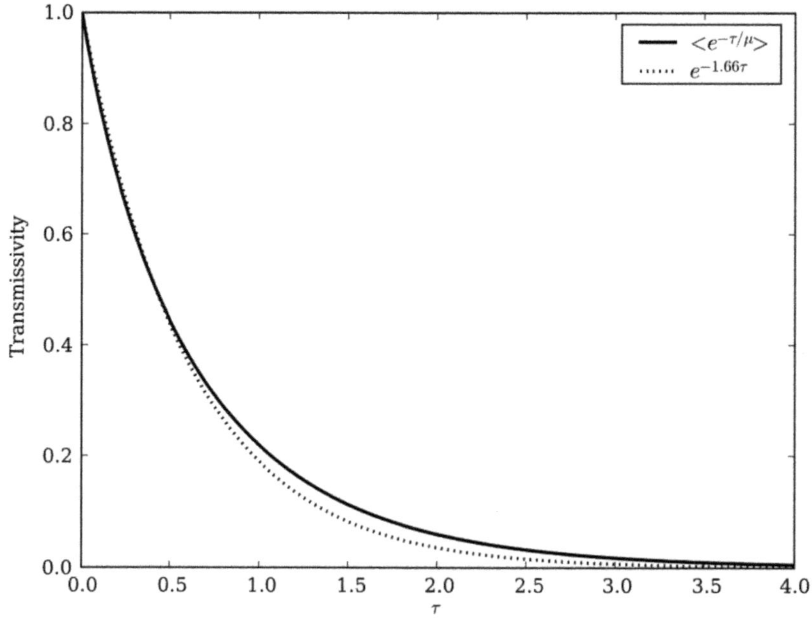

Figure 5.10. A comparison of the numerically computed angular average transmissivity $\langle e^{-\tau/\mu}\rangle$ (solid line) with the Elsasser diffusivity factor approximation $e^{-1.66\tau}$ (dashed).

$$\frac{dF^{\downarrow}}{d\tau} = F^{\downarrow} - \pi B. \tag{5.68}$$

Remember that the optical depth τ in (5.65)–(5.68) includes the 1.66 diffusivity factor.

5.7.3 Effective emission level

Let's consider now the radiation emitted by Earth to space. If the atmosphere is optically thick, radiation emitted by the surface will mostly be absorbed within the atmosphere. Thus, the radiation emitted to space will mostly originate at some level above the surface; the aim of this section is to determine that level.

In the plane-parallel diffuse approximation (5.65), the upward irradiance at the top of the atmosphere, also known as the *outgoing longwave radiation* (OLR), takes the form

$$F_{\infty}^{\uparrow} = \pi B_s e^{-\tau_\infty} + \int_0^{\infty} \pi B\, W(z)\, dz \tag{5.69}$$

where τ_∞ is the total optical depth of the atmosphere and we define the *weighting function*

$$W(z) = \frac{d}{dz} e^{-(\tau_\infty - \tau(z))} = \frac{d\tau}{dz} e^{-(\tau_\infty - \tau(z))} = \rho_a k_a\, e^{-(\tau_\infty - \tau(z))}, \tag{5.70}$$

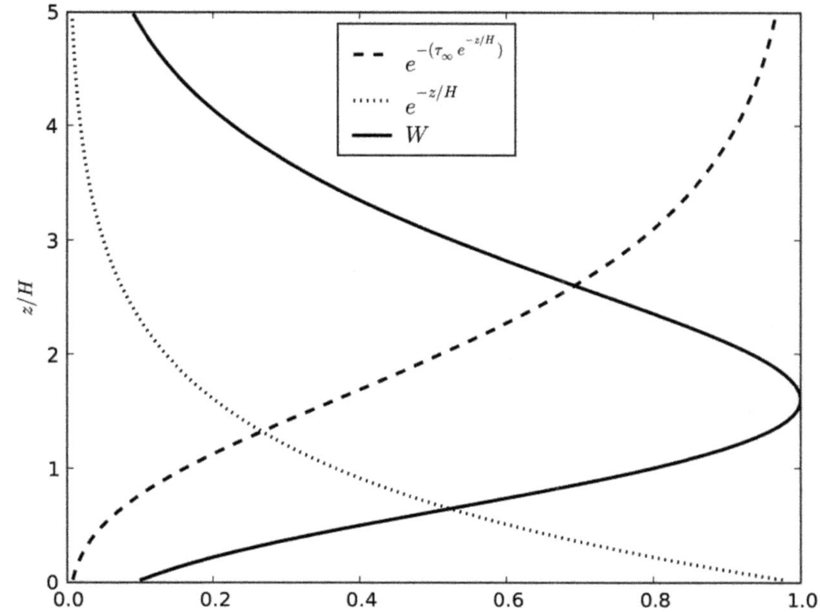

Figure 5.11. The weighting function W for $\tau_\infty = 5$.

where ρ_a is the density of absorbers. The weighting function determines how much each layer of the atmosphere contributes to the total outgoing radiation.

Now let's assume that (i) the atmosphere has a constant scale height, so that density varies as $\rho = \rho_s e^{-z/H}$; (ii) the absorbers are well mixed, so that the specific density $q = \rho_a/\rho$ is uniform; and (iii) pressure broadening does not play a big role, so k_a does not vary with height. With these assumptions,

$$\tau = qk_a \int_0^z \rho dz = \tau_\infty\left(1 - e^{-z/H}\right) = \tau_\infty\left(1 - \frac{p}{p_s}\right) \tag{5.71}$$

where in this case

$$\tau_\infty = Hq\rho_s k_a = \frac{qp_s k_a}{g} = u_\infty k_a \tag{5.72}$$

with u_∞ the total mass path of absorbers and

$$W(z) = \frac{\tau_\infty}{H} e^{-\left(\tau_\infty e^{-z/H}\right)} e^{-z/H}. \tag{5.73}$$

Figure 5.11 shows an example plot of $W(z)$ and its component terms. The salient feature of $W(z)$ is that it is strongly peaked, so there is a specific level (or neighbourhood) that contributes the major part of the outgoing radiation. This is called the *effective emission level* or *Chapman layer*. Physically, the effective emission level corresponds to the optimal trade-off between high density (which gives high emissivity) and little overlying atmosphere to permit the emitted radiation to escape. The location of the emission level is given by

$$\frac{dW}{dz} = \frac{W}{H}\left(\tau_\infty e^{-z/H} - 1\right) = \frac{W}{H}\left(\tau_\infty - \tau - 1\right) = 0 \tag{5.74}$$

which means that the emission level is at

$$\tau = \tau_\infty - 1. \tag{5.75}$$

Thus *most of the outgoing radiation comes from a level 1 optical thickness unit below the top of the atmosphere* (warning: in many texts, optical depth is defined to be 0 at the top of the atmosphere instead of at the surface as here; in those texts, the emission level is at $\tau = 1$). Using (5.71), the height of the emission level is

$$z_e = H \ln \tau_\infty \tag{5.76}$$

and the corresponding pressure level is

$$p_e = \frac{p_s}{\tau_\infty}. \tag{5.77}$$

5.7.4 The greenhouse effect

If the Earth had no atmosphere, what would be its surface temperature? This classic calculation goes as follows. Assume that the Earth is warmed to a uniform temperature T_e by the absorbed insolation and emits infrared radiation with emissivity 1. In steady state, the energy absorbed matches the energy emitted, so

$$\pi R_E^2 S_0(1 - \alpha) = 4\pi R_E^2 \sigma T_e^4 \tag{5.78}$$

where S_0 is the solar irradiance arriving on Earth—the so-called *solar constant*, though there are small variations over time—which has a value of $S_0 = 1367$ W m^{-2}; α is the albedo, taken to be 0.3, $R_E = 6.37 \times 10^6$ m is Earth's radius and σ is the Stefan–Boltzmann constant. Using these values

$$T_e = \left(\frac{S_0(1 - \alpha)}{4\sigma}\right)^{1/4} \approx 255\,\mathrm{K}. \tag{5.79}$$

T_e is known as the *emission temperature*. It is determined solely by the insolation and the planetary albedo. On Earth, T_e is much colder than the observed global-mean surface temperature of 15 °C or 288 K. The difference must be due to the atmosphere. The warming effect of the atmosphere, known as the *greenhouse effect*, is best understood as follows. The atmosphere is opaque in the longwave, which means that the mean effective emission level is lifted off the ground. The mean temperature at the emission level must be T_e in order for emission to match absorbed insolation. But the atmosphere has a positive lapse rate and so the temperature at the ground must be *greater* than T_e.

In general, an atmosphere must satisfy two conditions in order to provide a greenhouse effect: it must absorb radiation in the spectral range associated with black body radiation at temperature T_e and it must have a positive lapse rate. An atmosphere with a negative lapse rate (temperature *increasing* with height) will have a surface temperature *colder* than T_e. This is known as the *anti-greenhouse effect* and actually occurs on Earth in the polar regions during winter, as evidenced by figure 5.15(b).

5.8 Absorption and emission spectra of Earth's atmosphere

5.8.1 Absorption spectrum

The absorptivity spectrum by passage through the entire depth of atmosphere (without clouds) is shown in figure 5.12. Overall, the atmosphere is completely opaque in the ultraviolet ($\lambda < 0.3$ μm) and in the infrared (>4 μm), aside from a *spectral window* in the range 8–14 μm. The intermediate region, containing the visible, is much more transparent. Ultraviolet opaqueness is due to continuum absorption by ozone in the stratosphere. In the infrared, absorption is due to various species, as shown in figure 5.13.

The chief infrared absorbers, in decreasing order of importance, are

1. H_2O (water vapour), which absorbs all radiation in the ranges 5–8 μm and >20 μm. The region of weak absorption (8–17 μm) is called the *water vapour window*.
2. CO_2 (carbon dioxide), absorbing all radiation between 14 and 16 μm. Note that this band falls in the water vapour window, which gives it particular importance.
3. O_3 (ozone), better known as an absorber of ultraviolet (not shown in the figure), but also active in the infrared at about 10 μm, again falling in the water vapour window.

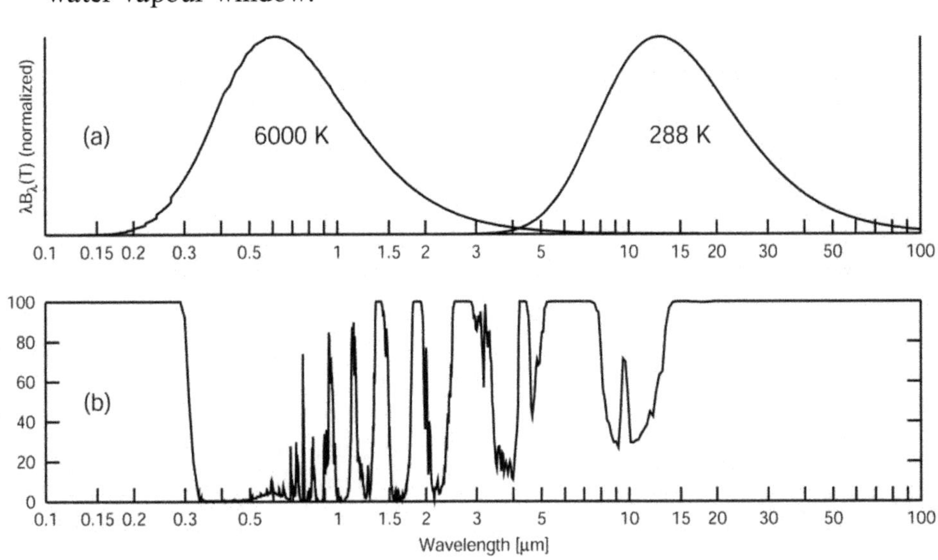

Figure 5.12. Whole-atmosphere clear-sky absorptivity compared with the Planck functions at the temperatures indicated. From Petty (2004).

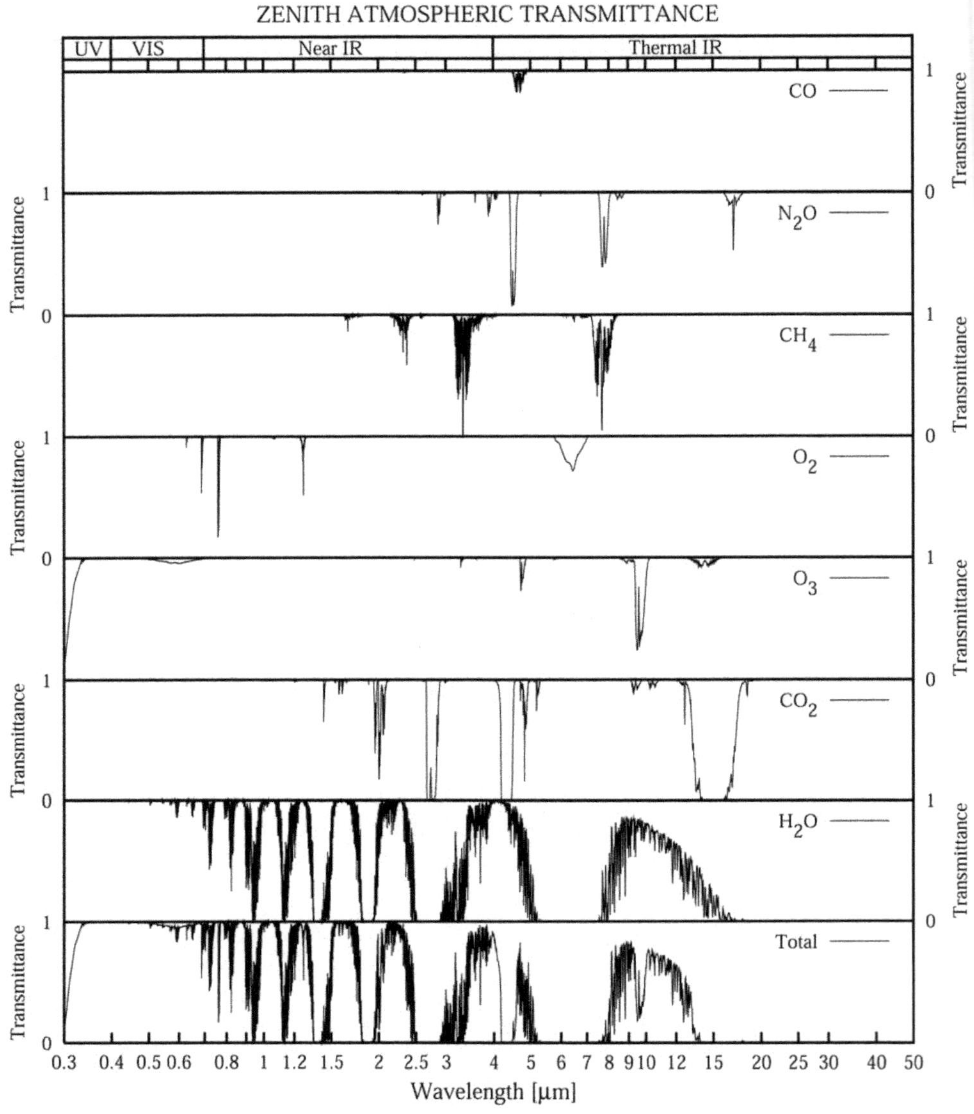

Figure 5.13. Breakdown of whole-atmosphere clear-sky absorptivity by contributing species. From Petty (2004).

4. N_2O (nitrous oxide) and CH_4 (methane) both have bands centred around 8 μm, near the shortwave end of the water vapour window. The absorption coefficient for these bands is high, but the mass path is small and so the absorptivity is not high. However, because these bands are unsaturated, their absorptivity is very sensitive to increasing mass path (see section 5.6.6): adding an extra molecule of N_2O or methane has a much greater effect than adding an extra molecule of CO_2.

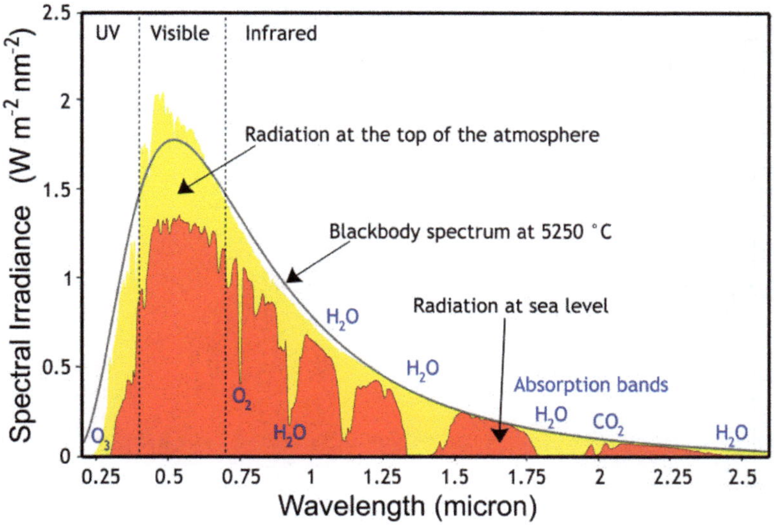

Figure 5.14. Spectra of solar radiation at the top of the atmosphere (yellow) and at the ground (red) under clear-sky conditions.

5.8.2 Absorption of solar radiation

The yellow area in figure 5.14 shows the spectrum of solar irradiance arriving at the top of the atmosphere. It follows a black body spectrum quite closely. A sunbeam loses about 20% of its power (under clear-sky conditions) on its way to the ground. Much of this is a continuum loss due to Rayleigh scattering: note the greater reduction at shorter wavelengths. There is also strong absorption in some bands, especially in the ultraviolet (ozone) and near infrared (water vapour). Solar radiation does heat the atmosphere directly, but most of its energy is deposited at the ground.

5.8.3 Emission spectrum

Earth emits as much infrared radiant energy to space as it receives in the form of solar radiation. The spectrum of outgoing longwave radiation, as observed by satellites in orbit around Earth, is shown in figure 5.15. Overlaid are black body spectra at various temperatures. In the jargon of remote sensing, the *brightness temperature* T_{B} is defined as the temperature a black body would need to have in order to emit the observed irradiance at a given wavelength

$$T_{\mathrm{B}} = \frac{hc}{k\lambda \ln\left(\dfrac{2\pi hc^2}{\lambda^5 I(\lambda)} + 1\right)}. \tag{5.80}$$

Physically, the brightness temperature is roughly the temperature at the effective emission level. The height of the emission level increases with optical depth. On Earth, the emission level is mostly within the troposphere, where temperature

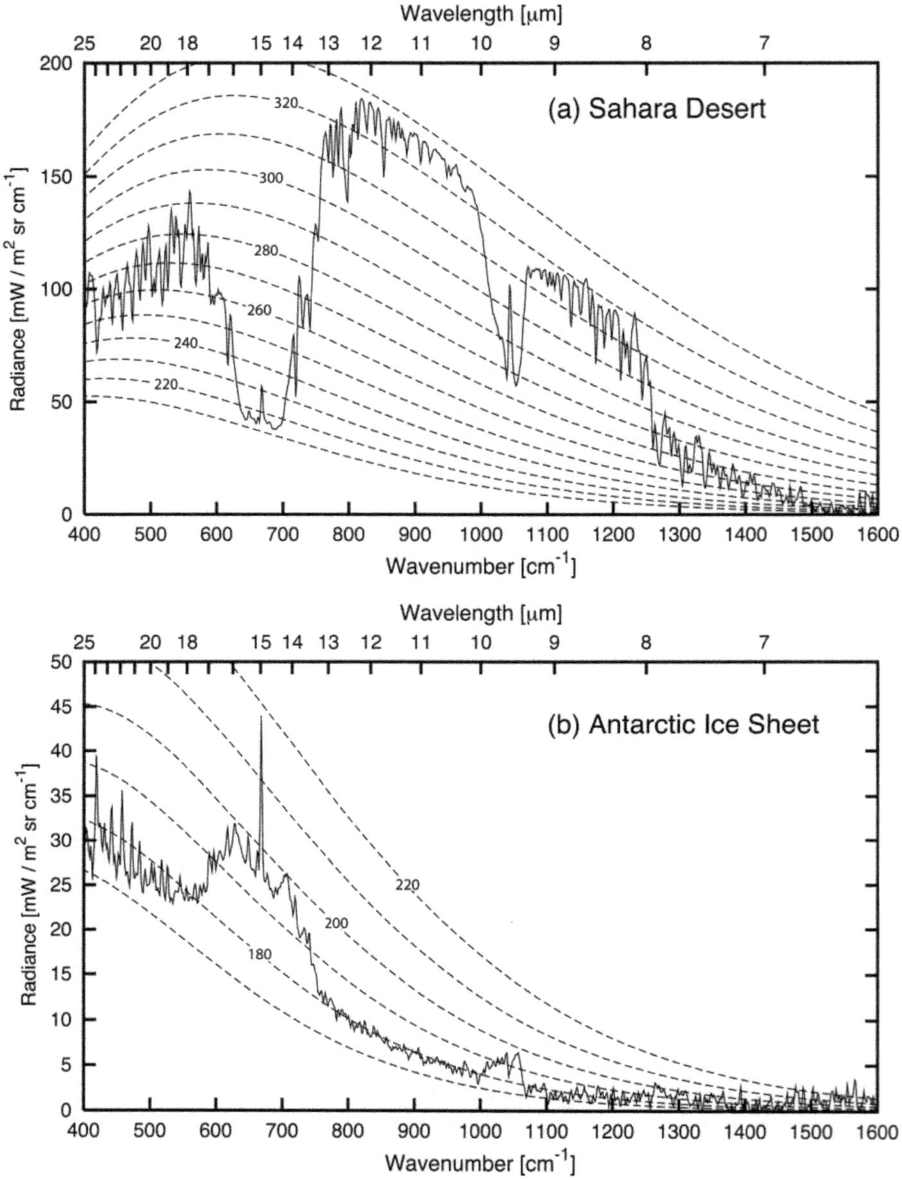

Figure 5.15. OLR spectra observed over (a) the Sahara and (b) the Antarctic ice sheet, overlaid by black body spectra at the temperatures indicated. From Petty (2004).

decreases with height. Thus, *the greater the optical depth, the lower the brightness temperature.*

In figure 5.15(a), the maximum T_B of about 330 K (57 °C) is in the water vapour window between 10 and 13 μm; this can be taken to be the temperature of the surface. The 15 μm CO_2 band has the minimum temperature (~215 K or −70 °C), which can be taken as the tropopause temperature (note how, at the very centre of the band, where optical depth is greatest, the brightness temperature is actually somewhat higher, this is because the emission level is in the stratosphere, which is warmer). The very opaque water vapour bands at wavelengths >20 μm and <8 μm emit at an intermediate temperature. The difference between the CO_2 and H_2O brightness temperatures is because water vapour is not well mixed: specific humidity decreases exponentially with height, so the emission level is lower than for CO_2.

Figure 5.15(b) presents a very different picture. Over the Antarctic ice sheet there is typically a strong temperature inversion near the surface and the surface is actually colder than the tropopause. Because of the very low temperatures, the atmosphere is also very dry, so there is very little water vapour absorption. As a result, the brightness temperature corresponds to the surface value of 180 K through most of the spectrum. The 15 μm CO_2 band, and also the 10 μm ozone band, show greater temperatures. The maximum brightness temperature of 220 K is the centre of the CO_2 band and comes from the stratosphere.

5.9 Some scattering effects

5.9.1 Radiative transfer in clouds

Until now, we have considered radiative transfer with absorption and emission but neglecting scattering, which is appropriate for transfer of longwave radiation through atmospheric molecules. Scattering cannot be neglected in clouds, however. Transfer of *solar* radiation through clouds involves scattering and absorption, but not emission. Consider the fate of an initially parallel beam of photons striking a cloud from above. If the single scatter albedo ω is small, absorption dominates over scattering: a few photons are deviated from their path, but mostly the beam is attenuated without scattering. As ω increases, more and more photons are scattered, creating a secondary, diffuse radiation stream. Some of the diffuse photons manage to find their way back up to the top of the cloud: the fraction of photons for which this happens is called the *cloud albedo*. Some of the diffuse photons exit through the bottom of the cloud, together with the remnants of the parallel beam.

Motivated by this qualitative picture, we formulate a heuristic model for radiative transfer within a cloud. The set-up is shown in figure 5.16. A parallel beam of sunlight with irradiance F_i is incident on the cloud from above. We assume the cloud has $\omega = 1$, i.e. it scatters but does not absorb solar radiation. The optical depth is given by

$$\tau = \int_0^z \rho_c k_s \mathrm{d}z \tag{5.81}$$

where ρ_c is the mass density of cloud condensate, also known as the *cloud water content*, and for convenience we take the z-axis pointing downward. As the beam enters the cloud, scattering occurs in both the forward and backward directions, setting up two *diffuse* streams of radiation, $F^\uparrow$ and $F^\downarrow$, in the up and down directions respectively. Both streams are assumed to be isotropic in their respective hemispheres, but we assume asymmetry between forward and backward scattering: the probability of backward scattering, denoted by b, will generally be different from the

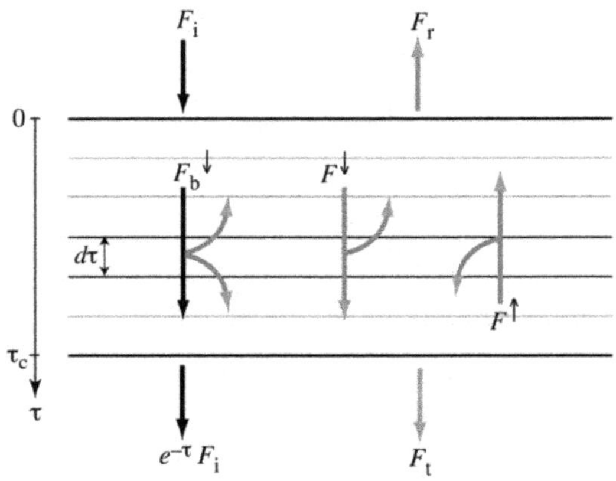

Figure 5.16. Schematic representation of radiative fluxes within a cloud. Black and grey arrows show beam and diffuse radiation, respectively. Curved arrows indicate the flux lost by backward scattering (beam radiation loses intensity by both backward and forward scattering).

probability of forward scattering $1 - b$. The diffuse streams exit through the top and bottom of the cloud with irradiances F_r and F_t respectively. We define the *reflectivity* or *albedo* as

$$\mathcal{R} = \frac{F_r}{F_i} \tag{5.82}$$

and the *diffuse transmissivity* as

$$\mathcal{T} = \frac{F_t}{F_i}. \tag{5.83}$$

With these assumptions, the radiative transfer equations take the form

$$\frac{dF_b^{\downarrow}}{d\tau} = -F_b^{\downarrow} \tag{5.84}$$

$$\frac{dF^{\downarrow}}{d\tau} = (1 - b)F_b^{\downarrow} - b\left(F^{\downarrow} - F^{\uparrow}\right) \tag{5.85}$$

$$\frac{dF^{\uparrow}}{d\tau} = -bF_b^{\downarrow} - b\left(F^{\downarrow} - F^{\uparrow}\right) \tag{5.86}$$

where $F_b^{\downarrow}$ is the beam irradiance within the cloud. Choosing $\tau = 0$ at the top of the cloud and $\tau = \tau_c$ at the bottom, the boundary conditions are

$$F_b^{\downarrow} = F_i, \quad F^{\downarrow} = 0, \quad F^{\uparrow} = F_r = \mathcal{R}F_i \quad \text{at } \tau = 0 \tag{5.87}$$

$$F^{\downarrow} = F_t = \mathcal{T}F_i, \quad F^{\uparrow} = 0 \quad \text{at } \tau = \tau_c. \tag{5.88}$$

This model can be derived as the plane-parallel, two-stream approximation to the radiative transfer equations in the case of scattering with no absorption. We will not go into the details except to note that the backward scattering probability b can be expressed as

$$b = \frac{1 - g}{2} \tag{5.89}$$

where

$$g = \frac{1}{2}\int_{-1}^{1} P(\mu)\mu \mathrm{d}\mu \tag{5.90}$$

is called the *asymmetry parameter* (here, P is the phase function, $\mu = \cos\theta$ and θ is the zenith angle between the directions of incoming and outgoing photons). Note that if the scattering is isotropic, $P = 1$, $g = 0$ and $f = 1/2$ (equal probabilities of forward and backward scattering).

To solve the model, we first note that for the beam radiation decays exponentially within the cloud:

$$F_{\mathrm{b}}^{\downarrow}(\tau) = F_{\mathrm{i}}\mathrm{e}^{-\tau}. \tag{5.91}$$

Subtracting (5.88) from (5.87) we find

$$\frac{\mathrm{d}}{\mathrm{d}\tau}\left(F^{\downarrow} - F^{\uparrow}\right) = F_{\mathrm{i}}\mathrm{e}^{-\tau} \tag{5.92}$$

which on integration, using the top boundary condition, yields

$$F^{\downarrow} - F^{\uparrow} = \left(1 - \mathcal{R} - \mathrm{e}^{-\tau}\right)F_{\mathrm{i}}. \tag{5.93}$$

Substituting into (5.87) and (5.88) then gives

$$F^{\downarrow}(\tau) = \left[1 - \mathrm{e}^{-\tau} - b(1 - \mathcal{R})\tau\right]F_{\mathrm{i}} \tag{5.94}$$

$$F^{\uparrow}(\tau) = [\mathcal{R} - b(1 - \mathcal{R})\tau]F_{\mathrm{i}}. \tag{5.95}$$

To complete the solution, we need expressions for $\mathcal{T}$ and $\mathcal{R}$ in terms of the total optical depth τ_{c}. Conservation of energy requires

$$\mathcal{T} + \mathcal{R} + \mathrm{e}^{-\tau_{\mathrm{c}}} = 1, \tag{5.96}$$

and evaluating (5.94), (5.95) at $\tau = \tau_{\mathrm{c}}$ gives

$$\mathcal{R} = \frac{b\tau_{\mathrm{c}}}{1 + b\tau_{\mathrm{c}}} \tag{5.97}$$

$$\mathcal{T} = \frac{1 - \left(1 + b\tau_{\mathrm{c}}\right)\mathrm{e}^{-\tau_{\mathrm{c}}}}{1 + b\tau_{\mathrm{c}}}. \tag{5.98}$$

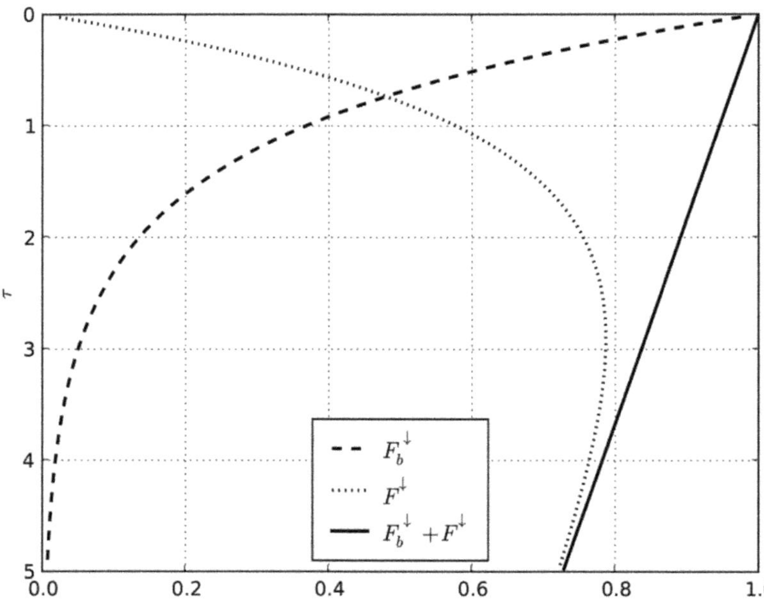

Figure 5.17. Beam irradiance $F_b^{\downarrow}$ and diffuse irradiance $F^{\downarrow}$ (normalized by the incident irradiance F_i) as a function of optical depth τ. The backscatter fraction $b = 0.075$, corresponding to asymmetry parameter $g = 0.85$, a reasonable value for 10 μm cloud droplets.

Finally, substituting in (5.94), (5.95) gives

$$F^{\downarrow} = \left[\frac{1 + b(\tau_c - \tau) - (1 + b\tau_c)e^{-\tau}}{1 + b\tau_c}\right]F_i \tag{5.99}$$

$$F^{\uparrow} = \left[\frac{b(\tau_c - \tau)}{1 + b\tau_c}\right]F_i. \tag{5.100}$$

These solutions are plotted in figure 5.17. Note that for $\tau_c > 1$, the transmitted irradiance will be mostly diffuse, and the light will cast no shadows. As $\tau_c \to \infty$, $\mathcal{T} \to 0$ and $\mathcal{R} \to 1$: a very thick cloud will be perfectly reflective.

How optically thick is a typical cloud? Following (5.27), the mass scattering coefficient for a droplet of radius r is

$$k_s = \frac{\pi r^2}{\rho_l 4\pi r^3/3} Q_s \tag{5.101}$$

where ρ_l is the density of liquid water and Q_s is the scattering efficiency. Given visible wavelengths of about 0.5 μm and typical cloud droplet sizes of about 10 μm, the scattering parameter is about 60, so we are deep in the Mie regime where $Q_s \approx 2$. This gives $k_s \approx 150\ \mathrm{m^2\ kg^{-1}}$. The typical cloud water content of stratocumulus clouds is $0.5\ \mathrm{g\ kg^{-1}}$ and the asymmetry parameter is never too far from 0.85, implying $b \approx 0.075$. With these values, the optical depth of a 400 m thick cloud deck will be about 30, giving an albedo of about 0.7. If the cloud is horizontally continuous, it will be pretty gloomy under this cloud.

5.9.2 Rain radar

Radar, which is short for *radio detection and ranging*, works on a simple and well known principle: send out a short pulse of radiation and carefully measure the time t taken for the light to bounce off an object and back to an antenna coaxial with the emitter. Then the distance d to the object is

$$d = \frac{ct}{2} \tag{5.102}$$

where c is the speed of light.

In meteorology, radar is used to detect *precipitation*. This requires a careful choice of wavelength: we want the radiation to pass unattenuated through clear air, aerosols and clouds, but backscatter strongly from raindrops. Recall from section 5.4 that the strength of scattering is controlled by the parameter $x = 2\pi r/\lambda$. There is negligible scattering for $x < 10^{-3}$. Cloud droplets and aerosols have radii $r < 100$ μm, while precipitation has $r > 1000$ μm. Thus, the wavelength should be chosen so that $r \sim 100$ μm makes $x \sim 10^{-3}$, which implies $\lambda \sim 10$ cm, in the microwave range.

Most meteorological radars in Europe use $\lambda = 5$ cm, the so-called C-band. This works fine in light and medium rain, but there is strong attenuation in strong rain so it is not suitable for severe weather. Since severe weather is much more frequent in North America, the US NEXRAD network uses the 10 cm S-band instead, which does not suffer attenuation but requires a larger antenna and is considerably more expensive.

Apart from showing *where* rain is, radar can also be used to estimate *how much* it is raining. This is done by measuring the intensity of the backscattered radiation. In general, a shaft of precipitation will contain drops of a range of sizes, each giving a distinct contribution to the backscattering. The total radiance backscattered by a slab of atmosphere of thickness ds is

$$\mathrm{d}I_\mathrm{b} = I\left(\int_0^\infty P(\pi)\, n(r)\, m(r)\, k_\mathrm{s}\, \mathrm{d}r\right)\mathrm{d}s = I\, P(\pi)\left(\int_0^\infty n(r)\, 2\pi r^2\, Q_\mathrm{s}\, \mathrm{d}r\right)\mathrm{d}s \tag{5.103}$$

where $P(\pi)$ is the value of the phase function for backward scattering ($\theta = \pi$), $n(r)\mathrm{d}r$ is the number density of drops with radius r, $m(r)$ is the corresponding drop mass, k_s is the mass scattering coefficient and Q_s is the scattering efficiency. Since we are in the Rayleigh regime (section 5.4.1), $P(\pi) = 3/2$ and $Q_\mathrm{s} \sim x^4$, so

$$\mathrm{d}I_\mathrm{b} \sim Z\mathrm{d}s \tag{5.104}$$

where

$$Z = \int_0^\infty n(r) r^6 \mathrm{d}r \tag{5.105}$$

is called the *reflectivity factor*.

Equation (5.104) shows that the backscattered intensity increases with both drop number density and with mean drop size. This is somewhat unfortunate, since it means that measurements of backscattered intensity are ambiguous: a high density of small drops may have the same Z as a smaller density of large drops. To remove the ambiguity, we need information about the size distribution $n(r)$. In a classic paper, Marshall and Palmer (1948) found empirically that

$$n(r) = N_0 \, \mathrm{e}^{-\Lambda r} \tag{5.106}$$

where N_0 and Λ are constants. Substituting into (5.105),

$$Z = N_0 \int_0^\infty r^6 \mathrm{e}^{-\Lambda r} \mathrm{d}r = N_0 \, \Lambda^{-7} \int_0^\infty z^6 \mathrm{e}^{-z} \mathrm{d}z = 6! \, N_0 \, \Lambda^{-7}. \tag{5.107}$$

We need to relate this to the rainfall rate R, defined as the mass of precipitation water falling on unit area in unit time. Using the by-now familiar argument that a flux is the product of a number density by a velocity, we find

$$R = \int_0^\infty v_t(r) n(r) m(r) \mathrm{d}r \tag{5.108}$$

where n is the particle size distribution as above, m is the mass of a particle and v_t is the terminal fall speed of precipitation. For precipitation-sized particles, the terminal fall speed can be expressed as:

$$v_t(r) = cr^d, \tag{5.109}$$

where c is a constant and the exponent d depends on the type of precipitation. Using (5.106), we have

$$R = \int_0^\infty cr^d r^3 N_0 \mathrm{e}^{-\Lambda r} \mathrm{d}r = C \, \Lambda^{-4-d}, \tag{5.110}$$

where all constants have been consolidated in C. Combining this with (5.107), we obtain

$$Z = aR^b \tag{5.111}$$

where a is a constant and $b = 7/(4 + d)$. For spherical drops, d is about 1/2 and $b = 14/9 = 1.56$. This compares well with the empirical *Marshall–Palmer relation*

$$Z = 200 \, R^{1.6} \tag{5.112}$$

where Z has units of $\mathrm{mm}^6 \, \mathrm{m}^{-3}$ and R has units of $\mathrm{mm} \, \mathrm{h}^{-1}$. This relation is quite accurate for moderate stratiform precipitation (i.e. the precipitation resulting from frontal lifting rather than moist convective instability). However, snowflakes have $d \approx 0$, giving $b \approx 2$. Also, the value of a is found to depend sensitively on the type of precipitation. In practice, it is necessary to first guess the type of precipitation being detected and then use the appropriate empirically derived coefficients in (5.111) to estimate the rain rate from Z. This introduces considerable uncertainty in the rain rate estimates.

Bibliography

Elsasser W M 1942 *Heat Transfer by Infrared Radiation in the Atmosphere* (*Harvard Meteorological Studies* vol 6) (Cambridge, MA: Harvard University Press)

Marshall J S and Palmer W M 1948 The distribution of raindrops with size *J. Atmos. Sci.* **5** 165–6

Petty G W 2004 *A First Course in Atmospheric Radiation* (Madison: WI, Sundog)

IOP Publishing

Physics of the Atmosphere

Rodrigo Caballero

Chapter 6

The atmospheric boundary layer

6.1 Scale separation and Reynolds averaging

Near the Earth's surface, atmospheric motion is generally *turbulent*, meaning it is unsteady, erratic and full of eddies of various sizes. This happens for two main reasons: one is that air in direct contact with the ground is slowed to a halt (a *no slip* boundary condition), introducing strong vertical *shear* (gradient) in the wind profile and resulting in dynamical instability which generates eddying motion; the other is that the surface may be warmer than overlying air, leading to static instability and consequent turbulent overturning.

Turbulent motion is too erratic and complex to describe in detail, either mathematically or numerically, and a central goal of turbulence theory is to express the *mean* effect of the turbulence on slow, large-scale components of the motion. Mathematically, if $U(\mathbf{x}, t)$ is the velocity field at a certain point and time, we write

$$U(\mathbf{x}, t) = \langle U \rangle + u'(\mathbf{x}, t) \equiv u + u' \tag{6.1}$$

where $u \equiv \mathbf{U}$ represents a *Reynolds average*—a spatial average over some neighbourhood of $\mathbf{x}$ and a time average over some interval centred on t. u' is the fast-varying, zero-mean part left over after averaging. This procedure assumes some sort of *scale separation*, i.e. it assumes that the physical processes controlling the evolution of u are different from and occur on typical space-time scales much larger than those controlling u', so that the two can be studied separately. Typically, u is taken to represent synoptic-scale phenomena, with time scales of several days, while u' represents turbulent/convective phenomena with spatial scales of up to a few kilometres and time scales of minutes or hours.

Applying decomposition (6.1) to all atmospheric variables, substituting into the equations of motion and applying $\langle \cdot \rangle$ to the equations, we obtain the *Reynolds-averaged equations of motion*:

$$\frac{\partial u}{\partial t} = -\frac{1}{\rho}\frac{\partial p}{\partial x} + fv - \frac{\partial \langle u'w' \rangle}{\partial z} \tag{6.2}$$

$$\frac{\partial v}{\partial t} = -\frac{1}{\rho}\frac{\partial p}{\partial y} - fu - \frac{\partial\langle v'w'\rangle}{\partial z} \tag{6.3}$$

$$\frac{\partial \theta}{\partial t} = J - \frac{\partial\langle \theta'w'\rangle}{\partial z} \tag{6.4}$$

where J is the diabatic heating and f is the Coriolis parameter. Several assumptions have been made to obtain the equations in this simple form: the *Boussinesq approximation* (essentially, density is taken to be constant), the *hydrostatic approximation* and the assumption that the turbulence is horizontally homogeneous, so that the horizontal derivatives of turbulence quantities drop out. Defining the *geostrophic wind*

$$fv_{\mathrm{g}} = \frac{1}{\rho}\frac{\partial p}{\partial x} \tag{6.5}$$

$$fu_{\mathrm{g}} = -\frac{1}{\rho}\frac{\partial p}{\partial y} \tag{6.6}$$

and assuming a steady state, the equations become

$$0 = f(v - v_{\mathrm{g}}) - \frac{\partial\langle u'w'\rangle}{\partial z} \tag{6.7}$$

$$0 = -f(u - u_{\mathrm{g}}) - \frac{\partial\langle v'w'\rangle}{\partial z} \tag{6.8}$$

$$0 = J - \frac{\partial\langle \theta'w'\rangle}{\partial z}. \tag{6.9}$$

The end result of Reynolds averaging is a set of equations for the slow components in which second-order averaged turbulence terms appear. A term such as

$$\rho\langle u'w'\rangle \tag{6.10}$$

has units of stress (N m^{-2}) and can be interpreted as a flux of u momentum; it is called a *Reynolds stress* or *eddy stress*. The term

$$-\frac{\partial\langle u'w'\rangle}{\partial z} \tag{6.11}$$

is the convergence of the eddy stress; it specifies the rate at which u momentum transported vertically by turbulence accumulates at each level, thereby accelerating the flow at that level.

6.2 Closure, mixing length and flux-gradient relations

The key point in the previous section is that, because the equations of motion are non-linear, the Reynolds-averaged equations describing the evolution of the slow components contain contributions from the turbulence. If we could somehow express the turbulent contributions as functions of the slow variables, then equations (6.2)–(6.4) would form a closed set, which could be solved to obtain the spatial structure and time evolution of the slow variables. This is the problem of *turbulence closure*.

Turbulence closure remains one of the great unsolved problems in the physical sciences. Though a fully rigorous mathematical solution is not known, considerable progress can be made using physical intuition supported by experiment. A key notion, developed by Prandtl in the 1920 s, is that of *turbulent mixing length*. The idea is to extend the classical kinetic-theory notion of transport by gas molecules to transport by macroscopic fluid elements. Between collisions, a gas molecule carries with it a fixed amount of momentum and energy; when it collides, it exchanges momentum and energy with surrounding molecules. Analogously, a fluid parcel carries with it a fixed quantity of momentum and energy until it 'collides', i.e. mixes with other parcels in the environment. Let us denote by ℓ the *mixing length*—the mean distance travelled by a fluid element without mixing, analogous to the mean free path of kinetic theory. Now supposing, for example, that u has non-zero vertical shear, the typical turbulent u fluctuation will be

$$u' \sim -\ell \frac{\partial u}{\partial z}. \tag{6.12}$$

If w is the typical turbulent vertical velocity, the vertical momentum flux is

$$\langle u'w' \rangle \sim -w\ell \frac{\partial u}{\partial z}. \tag{6.13}$$

This reasoning can be extended to turbulent transport of *any* quantity which is conserved by the motion, leading to the *flux-gradient relations*

$$\langle u'w' \rangle = -K_{\mathrm{m}} \frac{\partial u}{\partial z} \tag{6.14}$$

$$\langle v'w' \rangle = -K_{\mathrm{m}} \frac{\partial v}{\partial z} \tag{6.15}$$

$$\langle \theta'w' \rangle = -K_{\mathrm{h}} \frac{\partial \theta}{\partial z} \tag{6.16}$$

where K_{m} and K_{h} are called the *eddy diffusivities* for momentum and heat, respectively, and the minus sign indicates that the flux is directed *down* the mean gradient. Note that $K \sim w\ell$, but ℓ can be different for momentum and heat transport. We have just obtained the simplest example of a turbulence closure, which goes under the name of *K closure* or *K theory*.

6.3 The Ekman layer

Now let us use K closure in the steady-state equations of motion (6.7), (6.8). We will assume, heuristically, that K_{m} *is constant*. Then

$$f(v - v_g) = -K_m \frac{\partial^2 u}{\partial z^2} \tag{6.17}$$

$$f(u - u_g) = K_m \frac{\partial^2 v}{\partial z^2}. \tag{6.18}$$

Let us also assume that the geostrophic wind is constant in the vertical (i.e. barotropic) and align the x and y axes such that $v_g = 0$. It can then be shown that the solution to (6.17), (6.18) is

$$u = u_g\left(1 - e^{-\gamma z} \cos \gamma z\right) \tag{6.19}$$

$$v = u_g\, e^{-\gamma z} \sin \gamma z \tag{6.20}$$

where

$$\gamma = \sqrt{\frac{f}{2K_m}} \tag{6.21}$$

(this assumes we are in the northern hemisphere, so $f > 0$). This solution is the famous *Ekman spiral*, originally derived by Swedish oceanographer Ekman in 1905.

The Ekman spiral, shown in figure 6.1, starts from zero at the ground (as it should, given the no-slip boundary condition) and converges to the geostrophic wind for $z > \pi/\gamma$. In between, the wind is *to the left* of the geostrophic wind, i.e. it points towards low pressure. This cross-isobaric flow leads to mass convergence towards the centre of a low-pressure system (i.e. a cyclone). To conserve mass, boundary-layer convergence is balanced by ascent through the boundary-layer top and divergence in the free atmosphere. The overall effect is called *Ekman pumping*. Through the 'ice skater effect', divergence above the surface low tends to weaken the cyclonic vorticity and is an important damping mechanism for cyclones.

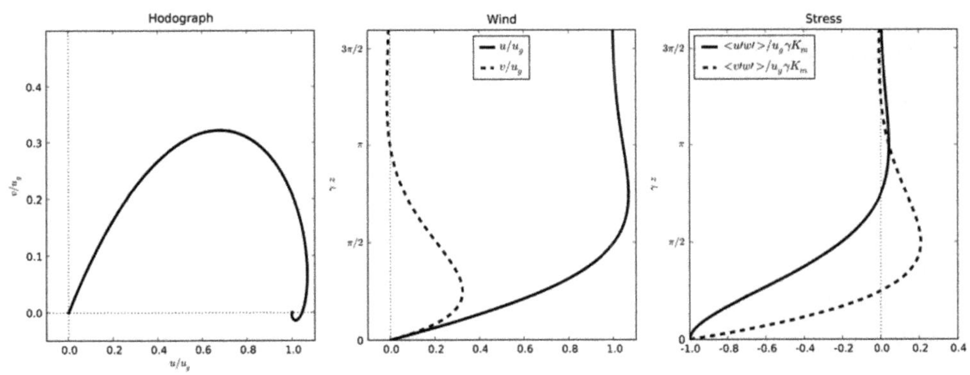

Figure 6.1. The Ekman spiral.

6.4 The surface layer

Ekman's solution fails near the surface, where the vertical shear is strong and the turbulent fluxes are strong enough that the Coriolis force can be neglected. This *surface layer* is typically ~100 m deep. The surface layer is thin enough that the eddy momentum flux does not deviate substantially from its surface value $\langle u'w' \rangle_0$. Let us define the *friction velocity*

$$u_\star^2 = \langle u'w' \rangle_0, \tag{6.22}$$

which is a characteristic turbulent velocity scale within the surface layer. Let us also assume that the layer is thin enough that the mixing length is limited by the distance z above the ground, so that

$$\ell = kz \tag{6.23}$$

where k is von Kármán's constant, which has an experimentally determined value of about 0.4. The flux–gradient relation is then

$$\frac{\partial u}{\partial z} = \frac{u_\star}{kz} \tag{6.24}$$

which can be integrated to yield

$$u = \frac{u_\star}{k} \ln\left(\frac{z}{z_0}\right). \tag{6.25}$$

The height z_0, called the *roughness length*, enters as a constant of integration; it is the nominal height at which the large-scale wind matches the no-slip boundary condition $u = 0$. The roughness length must be empirically determined; it is found to vary from a few millimetres above smooth water to several centimetres above grassy fields, to several metres over forests or urban areas. Because of the logarithmic dependence, the surface layer is often called the *log layer*.

6.5 Static stability and the Monin–Obukhov similarity

We have until now implicitly assumed that the atmosphere is neutrally stratified, so that KE derived from shear instability of the large-scale flow is entirely converted to eddy KE and successively lost to friction. However, we expect static stability to play a crucial role in turbulence. Under statically unstable conditions (e.g. when there is strong solar warming of the surface), there will be an extra source of KE for the turbulence, and the turbulent fluxes should be correspondingly stronger. When the mean flow is stable (e.g. due to night time cooling of the surface), air parcels will have to do work to move up or down, so turbulence will be weaker (another way of putting it is that some of the KE derived from shear instability will be converted to PE of the mean stratification).

The role of static stability can be quantified through the *flux Richardson number*

$$R_{\mathrm{f}} = \frac{g\langle\theta' w'\rangle/\theta}{\langle u'w'\rangle\,\partial u/\partial z} \tag{6.26}$$

where g is gravitational acceleration. It can be shown that the denominator is the rate of turbulent KE generation by shear instability (the *shear production* term), while the numerator is the rate of energy generation by static instability (the *buoyancy production* term). Assuming down-gradient momentum flux, the shear production term is always negative. When the ambient profile is statically stable, $\langle\theta' w'\rangle < 0$ (downward heat flux) and $R_{\mathrm{f}} > 0$, while unstable conditions imply $R_{\mathrm{f}} < 0$. For neutral stability there is no heat flux and $R_{\mathrm{f}} = 0$.

If $|R_{\mathrm{f}}| \ll 1$, then shear production dominates buoyancy (whatever the sign of the static stability) and we are close to the neutral case where the results of section 6.4 apply. Substituting (6.24) into (6.26), we find

$$R_{\mathrm{f}} = \frac{kg\langle\theta' w'\rangle/\theta}{u_\star^3}\,z \quad (|R_{\mathrm{f}}| \ll 1). \tag{6.27}$$

This means that $|R_{\mathrm{f}}|$ will increase with height. This will continue until $|R_{\mathrm{f}}| \approx 1$, where shear and buoyancy production are comparable. Above this level, buoyancy will dominate and (6.27) will not be valid. The transition from shear-dominated to buoyancy-dominated turbulence occurs at a height

$$L = \frac{u_\star^3}{kg\langle\theta' w'\rangle/\theta} \tag{6.28}$$

known as the *Monin–Obukhov length.*

How does static stability affect the wind profile? The answer is provided by *Monin–Obukhov similarity theory*, which is a generalization of (6.24):

$$\frac{kz}{u_\star}\frac{\partial u}{\partial z} = \Phi_{\mathrm{m}}(\zeta) \tag{6.29}$$

where

$$\zeta = z/L \tag{6.30}$$

is height adimensionalized by the Monin–Obukhov length and Φ_{m} is an unknown function that must be inferred empirically. Equation (6.30) cannot be mathematically derived from the equations of motion; it should be seen as a judicious guess. It is well supported by observations.

www.ingramcontent.com/pod-product-compliance
Ingram Content Group UK Ltd.
Pitfield, Milton Keynes, MK11 3LW, UK
UKHW051849100926
8556IPUK00005B/542

9 780750 310536